H.Y. Rasulov
S. S. Rakhimkhodjaev

Investigation and optimisation of the yarn tension on open-end weaving mills

H.Y. Rasulov
S. S. Rakhimkhodjaev

Investigation and optimisation of the yarn tension on open-end weaving mills

In the production of shirt fabrics

ScienciaScripts

Imprint
Any brand names and product names mentioned in this book are subject to trademark, brand or patent protection and are trademarks or registered trademarks of their respective holders. The use of brand names, product names, common names, trade names, product descriptions etc. even without a particular marking in this work is in no way to be construed to mean that such names may be regarded as unrestricted in respect of trademark and brand protection legislation and could thus be used by anyone.

Cover image: www.ingimage.com

This book is a translation from the original published under ISBN 978-620-7-48435-5.

Publisher:
Sciencia Scripts
is a trademark of
Dodo Books Indian Ocean Ltd. and OmniScriptum S.R.L publishing group

120 High Road, East Finchley, London, N2 9ED, United Kingdom
Str. Armeneasca 28/1, office 1, Chisinau MD-2012, Republic of Moldova, Europe
Managing Directors: Ieva Konstantinova, Victoria Ursu
info@omniscriptum.com

Printed at: see last page
ISBN: 978-620-8-36703-9

Contents

Annotation

The work is devoted to the research and optimisation of thread tension on the shuttleless weaving machines during the production of shirt fabrics. On the basis of the theory of wave propagation in elastic medium it is established that the cause of weft thread breaks is the impact of the brake foot on the weft thread, at the increase of the laying speed, the deformation and tension of the weft thread increase on average by 30% and at the increase of the thread friction coefficient on the surface of the laying mechanisms the tension and deformation of the weft thread decrease by 19%. A new system of weft braking and feeding is developed and tested. Regularities of change of the sinking string tension depending on the friction radius, friction angle, friction coefficient, sinking string stiffness and compensator position have been obtained. A stand and methodology for determining the friction coefficient in the compensator eye, depending on the friction radius, type and linear density of yarn, type of the compensator working surface are developed. Theoretical expert studies of warp thread tension for the existing and new system have been carried out, where the expediency of using the new system of warp thread tension regulation for the machine cycle, as the warp is triggered and during the machine start-stop period has been shown. The technological process of production on the weaving machine is investigated with the help of the mathematical method of rototable planning of the experiment of the second order. The geometrical interpretation of the mathematical model is studied by means of slices. The optimum technological parameters of shirt fabric production are determined, where the thread breakage is 0.3 breaks per 1 m of fabric at weft thread tension -10 cN and warp thread tension-20 cN. It is determined that technological parameters of production, structure and properties of shirt fabric depend on the properties used in weft yarns. With the increase of the stiffness modulus of weft yarns, the weft yarns processing decreases. With the use of cotton weft, the fabric has a high breaking load in the warp direction due to the high coefficient of friction between the warp and weft yarns. The breaking elongation in the warp direction for all fabric samples is different: the lowest in the fabric with cotton yarns in the weft is the highest in the fabric with nitron yarns in the weft.

Abstract. The work is devoted to the study and optimisation of yarns tension on shuttleless weaving looms in the production of shirt fabrics. Based on the theory of wave propagation in an elastic medium, it has been established that the cause of weft yarn breaks is the impact of the brake system on the weft yarn; with increasing filling speed, the deformation and tension of the weft yarn increase by an average of 30% and with an increase in the coefficient of friction of the thread on the surface of the filling insertion mechanisms tension and deformation of the weft yarns are reduced by 19%. A new braking and filling system has been developed and tested. The patterns of changes in the tension of the weft yarn depending on the radius of friction, the angle offriction, the coefficient offriction, the stiffness of the weft and the position of the compensator were obtained. A stand and method have been developed for determining the coefficient offriction in the eye of the compensator, depending on the radius of

friction, the type and linear density of the yarn, and the type of working surface of the compensator. Theoretical experimental studies of the warp thread tension for the existing and new systems were carried out, which showed the feasibility of using a new system for regulating the warp thread tension during the machine operating cycle, as the beam operates and during the machine start-stop period. The technological process of production on a weaving loom was studied using the mathematical method of second-order rotatable experiment planning. The geometric interpretation of the mathematical model is studied using slices. The optimal technological parameters for the production of shirt fabric have been determined, where the yarn breakage is 0.3 breaks per 1 m of fabric at a weft thread tension of -10 cN and a warp thread tension of -20 cN. It has been determined that the technological parameters of production, structure and properties of shirt fabric depend on the properties used in the weft of the threads. With an increase in the stiffness modulus of the weft threads, the processing of the weft threads decreases. With the use of cotton fabric has a high breaking load in the direction of the warp, due to the high coefficient of friction between the warp and weft threads. The elongation at break in the warp direction is different for all fabric samples: the smallest in fabric using cotton yarn in the weft, the highest in fabric using nitron threads in the weft.
Keywords: yarn, warp, weft, fabric, parameters, deformation, tension, coefficient, friction, system, properties, processing, modulus, stiffness, breakage.
Keywords: yarn, warp, weft, fabric, parameters, deformation, tension, coefficient,friction, system, properties, shrinkage, modulus, rigidity, breakage.

GENERAL CHARACTERISATION OF WORK

This is equally true for the textile industry, in particular for its most labour-intensive branch, which is weaving. The solution of this problem requires considerable research and development work aimed at improving weaving technology and loom mechanisms, as well as at optimising the parameters of shirt fabric production, taking into account the climatic conditions of the region. One of the main criteria for assessing fabric quality and weaving machine productivity is thread breakage, which is significantly influenced by the stability of warp and weft tension, which is determined, first of all, by the operation of the warp tempering and tensioning system and the weft feeding and braking system. The analysis of literature sources and the operating experience shows that these systems have a number of constructive drawbacks reducing technological and mechanical reliability of the process of warp and weft thread feeding and braking, which leads to the increase of warp and weft thread breaks and deterioration of the quality of the shirt fabrics produced. Therefore, the problem of optimising warp and weft tension is of undoubted interest for weaving production.

Purpose of the research task. Research and optimisation of warp and weft thread tension on machines with microspacers as one of the most effective ways to increase equipment productivity and improve the quality of shirt fabrics produced.

Objectives of the study:

- Study and optimisation of weft yarn tension in the process of shirt fabric formation;
- Study and optimisation of main yarn tension in the process of shirt fabric formation;
- Optimisation of shirt fabrics production technology;
- Technological and consumer research on shirt fabrics.

The object of the study is weaving machines with microspacers, shirt fabrics, yarn tension.

The subject of the research is methods and means of stabilisation of warp and weft thread tension, optimal parameters, technological and consumer properties of shirt fabrics.

Research Methods. The method of critical analysis of literary sources was used in the research process. When studying the tension of warp and weft yarns, methods of analytical geometry, theory of machine mechanisms and theoretical mechanics were used. In the experimental part the developed new systems of warp and weft tension, devices for determining the coefficient of friction of yarns against guides were used. The calculations were processed using the methods of computer technology.

The scientific novelty of the dissertation research consists in the following:

- New systems of warp feeding and braking, as well as warp release and tensioning systems are proposed;
- weft and warp yarn tensions for new systems were investigated and their regularities were obtained;
- mathematical model of thread breakage depending on the filling tension of warp

without costly machine adaptation. A disadvantage of the weaving process is the need to periodically create a relatively large space (shed) for the weft weaver; at a certain point in the development this limited the working speed. The negative role of this factor is reduced to a large extent with the multi-gape weaving system. The weaving machine has certain speed reserves for shed formation, especially with small doffing strokes. At the same time, the weft insertion speed is limited. The average speed of thread insertion in the shed is 15 m/s on shuttle looms - 15 m/s on micro shuttle and rapier looms - 30 m/s, on jet looms from 35 m/s to 40 m/s. The disadvantage of single shed weaving machines is that in the time interval corresponding to the laying angle, the weaver has to travel the full warp width. This disadvantage can be overcome by the following methods: introducing several weft threads at the same time (this method is realised on multigroove weaving machines); dividing the fabric into several parts which are joined by knitting (knitting and knitting machines) or increasing the speed of the weft insertion. Fabrics produced by the textile industry are classified according to the following characteristics: by fibre type (raw material composition); by the nature of fabric finishing; by the purpose of the fabric; by the weave of the fabric. By raw material composition fabrics are divided into the following groups - cotton, silk, woollen and linen, with more detailed division into subgroups within each group. In addition, fabrics by raw material composition are divided into homogeneous (base and weft from one kind of fibre), heterogeneous (base and weft from different kinds of fibre) mixed (base and weft are obtained from a mixture of different fibres in the process of spinning). According to the nature of finishing, fabrics are distinguished: harsh, which are not subjected to special treatments in finishing; raschlichten, which are subjected to soaking and washing to remove the schlichta; bleached, which have undergone the process of bleaching; smooth-dyed, which are dyed in one colour; printed, which are printed with coloured designs; mercerised, which are subjected to special treatment with caustic alkali; boiled, which are subjected to the process of boiling to remove sericin from the threads (silk fabrics); treated, which are subjected to special treatment to give the fabrics, low-shrinking, water-repellent, anti-decay and other properties; tufted, in which the tips of fibres are combed to the surface of the fabric or the tips of fibres are cut on special machines. By purpose fabrics can be: clothing - linen, dress, suit, coat; household - tablecloths, blankets, curtains, drapes, furniture, carpets; technical - filters, belts, tyres, oilcloths, linoleum, insulators. According to the types of weave fabrics are: simple, in which the patterns on the fabrics are obtained by the main weaves (canvas, twill, satin); finely patterned, in which the patterns on the fabrics are obtained on the basis of derivatives of the main and combined weaves; complex, which are formed by several systems of warp and weft yarns, i.e. fabrics with a specific look, i.e. fabrics with a specific type of weave.i.e. fabrics having a specific appearance - multilayered, openwork, etc.; large-patterned, in which the pattern on fabrics is formed by a combination of the above-mentioned weaves. Such a variety of raw material composition of fabrics is called a range of fabrics - cotton, linen, silk, woollen, etc. Taking into account the fact that price lists include about 3000 varieties of fabrics, which are called articles, having their own serial number, it will become obvious in the future to know the main features that will help the technologist to orientate in the assortment of fabrics. From all variety of assortment - 80 % are shirt fabrics having surface density not more than 100 gr./m^2 , the share of shirt fabrics produced by plain weave is 83 % in the basis of used twisted threads 30 %, also shirt fabrics 100 % are produced from threads of small linear density in the basis and weft. In summary it should be noted, taking into account climatic conditions of the region, it is necessary to produce shirt fabrics with surface density up to 100 gr./m^2 . It is also expedient to carry out the technology of shirt fabrics manufacturing without the process of sanding, by using twisted yarns of low linear density. Depending on their purpose fabrics should have appropriate consumer, physical, mechanical and hygienic properties determined by the type of fibrous material from which the fabric is made, its structure. Fabric structure is understood as the mutual arrangement of warp and weft yarns and their interrelations. The structure of fabrics is influenced by the following parameters: raw material composition, chosen taking into account the purpose of fabrics and the requirements that are imposed on them; warp and weft yarn diameters and their ratios, in which an increase in the diameter of yarns of one system increases the breaking load and elongation of the fabric of this system and the working out of yarns of the other system; fabric density by warp and weft, and their ratios, in which a change in the fabric density of one system of yarns causes a change in the technological parameters of fabric structure and properties; evaluation of the tension of production of the fabrics. The weaves having the smallest number of overlaps have a higher breaking load and yarn working out in the fabric, and the fabric working out is accompanied by a higher tension on the machine; technological parameters to which should be referred the tension of warp and weft yarns and their ratios, the size of the gap, the size and location of the shed. These parameters change the arrangement of threads in the fabric, hence determining the structure of the fabric (thickness, porosity, filling and stuffing of the fabric, etc.). In order to ensure a continuous process of fabric formation, it is necessary that the warp yarns have a certain tension. This tension is created by the warp release and warp tension mechanisms. The warp tension changes cyclically during each revolution of the main shaft of the machine. During weaving, the warp yarns are subjected to variable tensile forces, bending deformations and friction forces. During weaving, the warp is subjected to a cyclic return movement, which increases the abrasion effect on the warp. As the warp is travelling longitudinally at low speed on the loom, most of the above mentioned forces act repeatedly on the yarn. In order to withstand these dynamic forces, the warp yarn must be strong, elastic and abrasion resistant. In addition, it must be sufficiently smooth and even. It is known that cyclic warp tension leads to fatigue phenomena and increased yarn breakage. Therefore, in the process of fabric formation on the weaving machine, which is accompanied by cyclic changes in warp and fabric tension, much depends on the conditions created by the tempering and warp tension mechanisms. The set filling tension of the warp yarns for a fabric of a certain article must remain constant during the entire warp setting period. Only if this condition is fulfilled will the fabric have a uniform structure along its entire length. Therefore, the release and tensioning mechanisms with which the filling tension is set must not only ensure that the filling tension is constant in value, but also keep it constant during the entire warp setting period. The development of yarn fatigue is influenced not only by the maximum and minimum warp tension in the machine cycle, but also by the relative duration of the maximum and minimum loads. The lower the yarn fatigue limit, the longer the yarns are subjected to the maximum load during the machine cycle. The greatest efficiency of the weaving machine is achieved with lower warp tension. If the average tension is the same, the machine system with a smaller amplitude of warp tension fluctuations is more productive. When the filling tension increases, the elastic properties of the yarns decrease, their abrasion in the eyes of the tack and reed teeth increases, which leads to increased breakage, lower labour and equipment productivity. At low filling tension is sticking threads at the opening of the shed, and this makes it difficult to lay weft. During weft surfing at a sharp increase in warp tension on STB machines, there is a significant deformation of the moving system links of the scalp and the weaving warp, and after surfing there are torsional vibrations of the weaving warp, as a result of which the warp yarns experience additional jerks. Therefore, one of the measures to reduce warp breakage is the elimination of warp oscillations. The period of unsteady operation differs significantly from the conditions of fabric formation in the weaving machine's steady operation and, as a consequence, leads to the production of fabric with a weft density that differs from the specified weft density, i.e. to the occurrence of a fabric defect - starting stripes.

In the production of some articles of fabrics, 20-25% of the total number of defects are start-up streaks, and the size of the defect increases with increasing speed of the main shaft of the machine. Here are a number of recommendations to minimise starting fringes:

- reduction of tension in the filling system at the moments of machine standstill; here we should recommend technological measures - stopping the machines in the scoring position and working with a low filling tension, which is accompanied by a large size of the surfacing strip, which is less sensitive to the movements of the edging; design - the development of a mechanism that unloads the filling system automatically at the moment of machine standstill from most of the existing tension and automatically restores it at the moment of starting the machine after standstill;
- correct selection of existing and introduction of new special elements in different machine zones, allowing to change the pattern of the fabric edge movement during machine standstill;
- changing the filling patterns of special machines in order to reduce the length of fabric in the filling;
- change of yarn properties towards reduction of rheological properties, correct choice of modes of preparation operations for weaving; - creation of electrical and

electromechanical devices for controlling the position of the fabric edge during idle time and its return to the initial position at the moment of machine start-up while maintaining the initial level of filling tension.

The weft insertion process on the STB needleless weaving machines consists of winding up and transporting (inserting) a section of yarn into the shed by means of a micro-weaver accelerated by a torsion-type warmer. Depending on the filling width, there are 13 to 17 grommets used on the machine. The normal operation of the process in accordance with the technological requirements is ensured by the coordinated operation of the following mechanisms: weft brake, yarn availability controller, compensator, centring flap, grommet brake, etc. The length of the yarn to be transported ranges from 180 to 340 cm depending on the filling width of the machine. The machine is supplied with weft by fixed conical cross-winding bobbins located on the left side of the machine. On the STB machines, the thread is wound off the feeding bobbin only during the weft laying period, and during the rest of the machine operation the thread has a possibility to slacken between the bobbin and the guide eye of the cylinder limiter. The weft insertion speed through the shed depends on the linear density of the filling yarn and the twist angle of the torsion roller and varies from 19 to 26 metres per second. The number of counts reaches 200 to 300 per minute. The weft tension during the run-up period of the weaver reaches 80 % of the permissible breaking tension, and the average tension value during the free flight period of the weaver is 25 % of the permissible tension. The reasons for the change of tension values in the zones are linked to the cyclograms of weft mechanisms operation and conditions of yarn winding off the feeding bundle. The conditions of yarn winding off the bundle are a very important factor affecting the productivity of the machine. In order to reduce the weft breakage during shedding through the shed, it is recommended to use weft storages, to improve the yarn quality and its strength characteristics, to reduce the shedding speed. The greatest influence on the tension is the weft brake, and the brake with impact. In order to eliminate the impact when braking the yarn, it is proposed to feed STB machines with weft yarn from a rotating bobbin, which is driven into rotation by the winding yarn. The bobbin rotates in the opposite direction to the rotation of the ballooning yarn, which is wound at a speed of 1200 m/min, the angular speed of the rotating bobbin, which is driven into rotation by the winding yarn, does not exceed the angular speed of the winding weft yarn. In order to reduce the "impact" on the weft yarn when braking it, instead of a brake foot, a rotating roller made of a light material (ceramic) with a rim made of a hard material (porcelain ceramic) is used in the construction. In another design of weft brake, two elastic rings are provided instead of a brake foot. The pressure on the thread of each of the two rings is twice less than the pressure of the foot in the brakes of the factory design, which reduces the jerk in the tension of the laid thread at the time of passing under the foot brake knots or thickenings. To eliminate the impact on the thread, the weft brake has a damping foot in the new design of the thread tensioner. In the revised design, the brake band is directly connected to two springs, the pre-tensioning of

which is changed by an adjusting screw. The presence of several springs and the absence of a link (rocker arm) reduces the possibility of natural oscillations in the brake. The damping foot reduces the pre-tensioning force of the spring, and is particularly effective when processing yarns of low linear density. The question of the mechanics of a material deformable yarn constrained by friction is very widely covered in the literature. As it is known, L. Euler worked a lot in this direction, where for the first time he established the relation between the tension of the leading and the driven parts of a flexible link sliding on the surface of a cylinder. L. Euler derived his formula for a cylinder installed horizontally, and the flexible thread embraces the surface of this cylinder, located on it in a plane parallel to the cylinder guide. The ends of the thread hang down from the cylinder and are loaded by forces T_0 and T. The cylinder is stationary, the thread slides along its surface. The thread is weightless and non-stretchable and has perfect flexibility.
Under these conditions, he obtained the ratio:

$$T = T_o e^{k\varphi}$$

where: φ – *is the* girth angle, *k is the* friction coefficient of the flexible bond on the cylinder surface.
The questions of mechanics of a weighted deformable flexible thread on a plane and other forms of guides are considered. *A* thread of length *l* sliding on a plane has a tension from its own weight *q*

$$T_{пл} = q \cdot l \cdot \kappa = q \cdot r \cdot \varphi \cdot \kappa$$

A thread sliding along the circumference has a tension at the circumference arc $l = r \cdot \varphi$, :

$$T_{кр} = \frac{2q \cdot r \cdot \kappa}{1+\kappa^2}\left(e^{\kappa\varphi} + \frac{1-\kappa^2}{2\kappa} \cdot \sin\varphi - \cos\varphi\right)$$

The presented formulas determine the thread tension as a function of thread mass, friction coefficient and guide radius. The above formulas do not take into account the stiffness of filaments on the friction surface, as this parameter takes into account the genus and type of filaments, linear plane of filaments and elastic properties of filaments. Therefore, it is reasonable to study the tension of filaments on the basis of taking into account the coefficient of stiffness of filaments. To measure the coefficient of friction of rest and motion, a device is proposed, which contains a steel cylinder in contact with a textile thread, one end of which has a constant tension and the other end has a variable tension. The proposed device is inconvenient in maintenance and does not provide efficiency and accuracy of measurements, due to the lack of dynamometer and removable materials under study, as all operations are carried out manually. In conclusion, it can be noted that the existing weft tensioning systems have a number of drawbacks that reduce the reliability of the weft laying process. Modernisation of weft tensioning systems by the authors have not found application due to the complexity of designs and low efficiency in operation. The dynamics of the weft laying process is insufficiently studied. Taking into account the above-mentioned, the following tasks were set in the process of work performance - to investigate and stabilise warp and weft tension in the process of fabric formation.

The second chapter contains "Investigations of weft thread tension on machines with microspacers for shirt fabrics". On weaving machines with microspacers, weft weavers are used instead of a shuttle carrying a weft pack for weft insertion. The small size of the weft weavers allows a significant increase in machine speed. The weft yarns are wound from fixed bobbins, the number of which is determined by the weft colour. In addition to the weft weavers, the following mechanisms are involved in the weft insertion process on machines with micro-weft weavers: weft weaver spring expander, weft weaver lifter, weft thread returner, weft thread returner spring expander, shear selvedge mechanisms, receiving box mechanisms, weft weaver braking mechanism, weft weaver landing controller, weft weaver return mechanism, weft weaver stacker, weft weaver conveyor feeding weft weavers from the receiving box to the weft weaving box. When the weft stacker takes its home position in the box, the hook of the weft stacker spring expander mechanism enters the weft stacker. The weft weaver is then fed to the flight line by the lifting mechanism. According to the preset programme, the colour changing mechanism moves the required weft catcher with weft yarn to the weft weaver flight line. After the weft weavers have been lifted onto the flight line, the lifting mechanism stops and the hook opens, travelling in the opposite direction. The thread catcher grips release the weft yarn and the weft yarn is transferred from the thread catcher to the weft weaver. After capturing the weft yarn weaver weft combat mechanism accelerates to a speed of 20-24 m / s and the weft weaver makes a free flight in the shed batana towards the receiving box, as a result of which the weft yarn is laid. In the receiving box, the weft weaver is stopped by the braking mechanism, and the correctness of its landing is checked by the controller. The weft weaver is then fed to the fabric selvedge by a return mechanism and positioned so as to ensure the minimum length of the weft yarn tip to be bent by the needle of the selvedge forming mechanism. In addition, the return mechanism ensures proper operation of the grommet spring expander mechanism and the grommet ejector mechanism. The grommet spring expander mechanism is designed to release the weft yarn caught by the grommet jaw clamps and push the weft grommets out of the take-up box into the guide slot. At the moment when the weft weaver releases the yarn, the opening mechanism is fed into the guide slot by the laying mechanism, then the weft weavers from the guide slot are fed to the conveyor, which transports them to the weft weaving box. Fig.2.1 shows the technological scheme of weft threading of the machine with microlayers. The weft thread 2 coming off the bobbin 1 envelopes the system of thread guides 3, 5, 6 made in the form of wear-resistant porcelain eyes, weft brake 4 is clamped by the grips of weft returner 7 and before trolling is transferred to the weaver 8. The latter puts the weft through the shed formed by the warp threads. In the receiving box, the weft weaver is slowed down and fed by the weft returner in the opposite direction - to the selvedge. Compensator 5 selects excess threads from the shed, after which the thread is centred by the sash and held by the grippers, then the weft is released from the thread and pushed onto the conveyor, and simultaneously the left part of the thread is cut off with scissors.

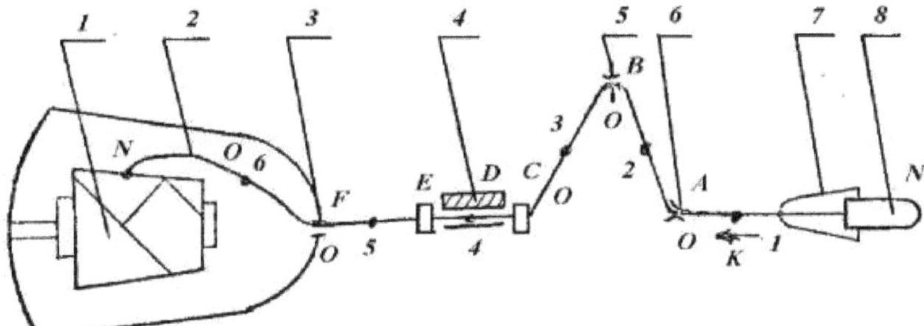

Fig. 2.1. Technological scheme of duck filling of the machine with microspacers.

According to the cycle diagram, the thread is inserted into the shed at 140° -295° , i.e. 155^{0} turns of the machine main shaft. Subsequently, the thread is pulled out of the shed by the weft compensator in the direction of the battle box, i.e. in the opposite direction to the shedding direction. The inductive brake becomes effective at a main shaft rotation angle of 180^{0} . From the oscillogram (Fig. 2.2.) it can be seen that at the moment of the beginning of braking the tension increases to the maximum, and at this moment there is an impact on the weft thread as it moves. This pre-braking is necessary to eliminate the excess weft in the fabric. The weft brake of these machines works according to a rigid cyclogram, i.e. the braking process takes place at a certain position of the main shaft, and the weft weaver flies into the take-up box and is braked at different times. The moment when the weft weaver flies in depends on the filling width of the machine, the speed of the main shaft, the initial flight speed, the linear density of the weft yarn, changes as the bobbin is triggered and for a number of other reasons. If the weaver flies into the receiving box and stops before the weft thread brakes, the thread will inevitably be thrown into the shed and the compensator will not be able to pick out the excess thread, which will lead to the machine stoppage or the production of defective fabrics. On STB machines, the thread is pre-braked so that the weaving process is not interrupted at almost any point in the insertion of the weaver. For a more detailed analysis of the influence of various factors on the weft tension, the oscillogram shown in Fig. 2.2. is divided into zones. Each zone in the weft insertion cycle corresponds to a change in one or another of the influencing factors.

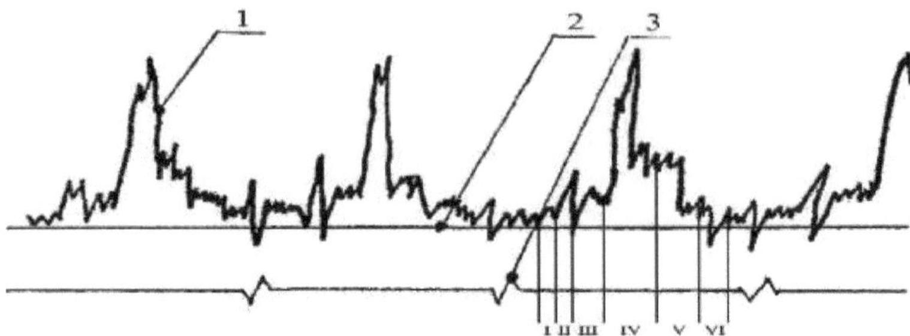

Fig. 2.2. Oscillogram of weft yarn tension

The first zone (see Fig. 2.2, I) starts at the moment of weft weaver acceleration (140° of machine main shaft rotation). The zone ends when the weft weaver reaches its maximum speed. Together with the weft weaver, a section of the yarn also moves. During a very short time - 0.006 s - the speed of the weaver with the yarn changes from zero to 19-26 m/sec. The tension in this section changes from zero to some value, and the increase is almost instantaneous, which corresponds to the first peak on the oscillogram. The tension depends on the acceleration of the grommet acceleration, which in its turn is a derivative of the torsion roller winding angle. The thread moving behind the weaver is selected from a free lying section prepared by the compensator. The second zone (see Fig. 2.2, II) corresponds to the beginning of the free flight of the grommet with the weft thread following it through the shed. It starts at the moment of thread tension decrease, when the thread reserve created by the compensator has not yet been selected. Zone II ends at the top of the second peak on the tension curve, which corresponds to the direct winding of the thread from the bobbin. The tension in this zone depends on the compensator advance angle, on the linear density of the weft yarn, on the weight of the yarn in the slack section, and on the homogeneity of the yarn structure. The third zone (see Fig. 2.2, III) is characterised by the fact that the tension in this zone stabilises after a certain decrease and remains uniform throughout the entire interval. The value of tension in this area depends on the conditions of winding from the feeding bobbin, the friction of the thread on the guiding elements, and the angles of thread girth. This period lasts until the weft is braked by the weft brake foot. At the moment of the beginning of braking the tension increases sharply, which on the oscillogram on the border of the third and fourth zones corresponds to the peak, the value of which depends on the weft number, the speed of the weft weaver, the condition of winding from the feeding bobbin. The fourth zone (see Fig. 2.2, IV) is characterised by some tension increase, which is created by the action of the brake and controller on the weft thread. The tension in this zone is the sum of the tension in the third zone plus the braking action of the brake foot and the weft controller stylus. This moment is quite stressful for the weft yarn, as the yarn is under the action of the mechanisms for a long time. For this zone is particularly important to acquire the conditions of winding off the bobbin, due to the fact that the length of the laid thread, moving behind the shuttle, reaches in this zone the maximum value - 220 cm and the maximum weight. The tension that the leading end of the weft yarn receives significantly reduces the speed of the weft weaver. For yarns with a high line density, the speed reduction is particularly high, reaching up to 1.2 m/s per metre of filling width. The fifth zone (see Fig. 2.2, V) of the weft insertion cycle through the shed starts from the moment the weaver enters under the first brake of the take-up box (beginning of the weaver braking). This moment is characterised by a sharp decrease in the speed of the weaver and the initial section of the moving yarn. But since the thread is flexible, relatively not rigid, and remains in motion when moving from the left box to the right, under the action of inertia forces it continues straight-line movement, which creates some excess thread in the shed. On the thread tension curve, this moment corresponds to the approach of tension to the zero line. Under certain conditions, the thread reserve in the shed can reach such a length that the compensator cannot pick it up. As a result, a loop is formed in the shed, which is a violation of the weaving process. In normal weft insertion, the excess yarn from the shed must be removed by the compensator before the weft weaver returns. The yarn tension in this zone depends on the tension in the previous zones, the shuttle arrival speed, yarn line density, correct setting of the weft brake, etc. The yarn tension in this zone depends on the yarn tension in the previous zones. The sixth zone on the tension curve (see Fig. 2.2, VI) starts with the beginning of the weaver return. It is characterised by tension constancy over the whole interval with very slight fluctuations at certain moments. In zone VI, after the spacer returns, the centred weft yarn, which has a certain tension required for normal fabric formation, is gripped by the weft return jaws. The yarn section, which is in the shed and held by the selvedge-former grippers, is cut off from the other section winding off the bobbin. At the same time, the end of the weft yarn is brought back to the initial position by the weft returner for transferring it to the weaver. Analytical studies of weft laying on machines with microspacers have also been carried out. The main causes of weft breakage and instability of the technological process of weft laying are jump-like (instantaneous) changes and deformations in the yarn sections. Therefore, it is expedient to develop analytical methods for calculating changes in the yarn motion parameters in interaction with auxiliary laying mechanisms and to investigate the causes of weft thread breakage in weaving. The theory of wave propagation in elastic medium is used as the main research method. Let at some moment t of time the weaver will come into motion. We will consider the action of the combat mechanism as a blow made on a resting and undeformed, initially flexible thread. The action of impact on the thread will propagate as a longitudinal wave travelling with velocity $a = \sqrt{\frac{E}{\rho_0}}$ where E is the dynamic modulus; ρ_0- is the density of the undeformed thread.

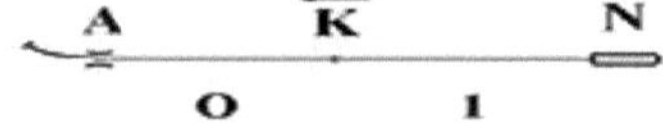

Figure 2.3. Schematic of the longitudinal wave motion at the NA section

Since the mass of the spacer is sufficiently large and the motion starts instantaneously, we can assume that in some time interval to < t < ti, the perturbed region of the filament I is the region of constant parameters. In this case, the velocity of the filament particle in region I is equal to the velocity of the spacer (since region I is a region of constant parameters). The thread tension is determined from the following equations

$$x_1' = -a\varepsilon_1 \quad (2.1) \qquad T_1 = \varepsilon_1 E = \rho_0 a^2 \varepsilon_1 \quad (2.2)$$

where *x' is the* tangent to the filament component of the part velocity in region I; e1 is the relative strain; T1 is the filament tension.

Equation (2.1) is derived from the law of conservation of quantity of motion when the yarn element under consideration passes the longitudinal wave front K (Fig. 2.3.), and equation (2.2) expresses Hooke's law for a linear elastic yarn. Thus, the tension and relative deformation of region I of the weft yarn are determined from the following equations

$$\varepsilon_1 = \frac{x_1'}{a} = \frac{U_1}{a} \quad (2.3) \qquad T_1 = \rho_0 \cdot a^2 \cdot \varepsilon_1 = \rho_0 \cdot a \cdot U_1 \quad (2.4)$$

Calculations were carried out for a thread with a closer to real textile threads a = 800 m/sec, $\rho_0 = 1{,}25 \cdot 10^{-3}$. After the time $\frac{NA}{a} = t_1 > t_0$ the action of the impact of the combat mechanism will pass to the thread conductor at the point Ai, where NA is the length of the horizontal part at the initial moment of time. At t >4, the longitudinal wave K passes to region 2 of the filament (Fig. 2.1., Fig. 2.4.).

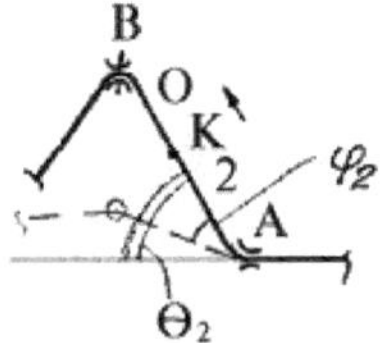

Fig. 2.4. Schematic of the longitudinal wave motion on the AB section

The velocity of motion of the particles of region 2 of the filament is determined from the law of conservation of the quantity of motion at the front of the longitudinal wave K. The law of the quantity of motion written in projections on the axes OX and OY (Fig. 2.5.) is reduced to the following equations

$$X_2' - X_0' = a(\varepsilon_2 - \varepsilon_0)\cos\theta_2 \cos\varphi_2 \qquad Y_2' - Y_0' = a(\varepsilon_2 - \varepsilon_0)\sin\theta_2 \sin\varphi_2$$

Since it is assumed that the thread is initially not deformed and is in a state of absolute rest, it is obvious $\varepsilon_0 = X_0' = 0$ and therefore the last equations can be written in the following form

$$X_2' = a\varepsilon_2 \cos\theta_2 \cos\varphi_2 \quad (2.5) \qquad Y_2' = a\varepsilon_2 \sin\theta_2 \sin\varphi_2 \quad (2.6)$$

To determine the tension Tg and relative strain £2 we have the following equations

$$T_2 = \rho_0 a^2 \varepsilon_2 \quad (2.7) \qquad T_1 = T_2 e^{f_2\theta_2} \quad (2.8)$$

where f_2 *is the* coefficient of friction at point A.

Equations (2.5)-(2.8) form a system of relative unknowns X_2', Y_2', T_2, ε. Excluding Ti and T_2 from equation (2.7) and (2.8)

let's find $\varepsilon_2 = \varepsilon_1 e^{-f_2\theta_2}$ (2.9)

since the deformation £1 is known from the solution of the problem for equation (2.9) serves to determine the deformation £2. The velocities of particles *X \, Y J are* determined from equation (2.5), (2.6) respectively. The section of the thread BC (Fig. Fig. 2.1., Fig. 2.5.) is divided into two regions. Region 3 (Fig. 2.1., Fig. 2.5.) is perturbed by the impact force of the laying mechanism, and region 1 is the
by the region of absolute rest. In region 3 and at the front K we have the following equations $X_3' = a\varepsilon_3 \cos\theta_3 \cos\varphi_3$ (2.10)

$Y_3' = a\varepsilon_3 \sin\theta_3 \sin\varphi_3$ (2.11) $T_3 = \rho_0 a^2 \varepsilon_3$ (2.12)

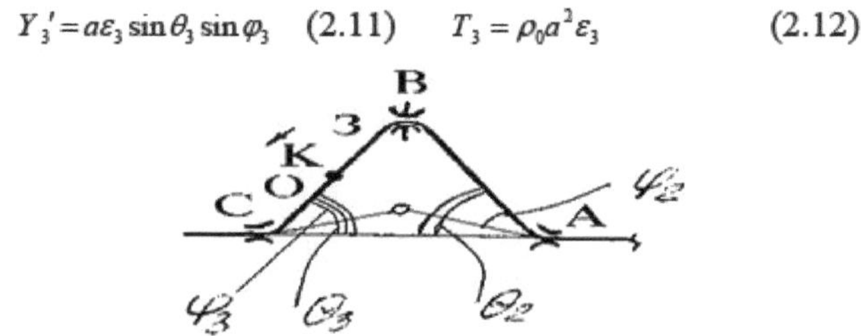

Fig. 2.5. Schematic of longitudinal wave motion on the BC section

The Euler equation at the (compensator) surface takes the following form $T_2 = T_3 e^{f_3(\theta_2+\theta_3)}$ (2.13), where f_3 is the dynamic coefficient of friction at the compensator face. Equations (2.10)-(2.13) form a system of relative unknowns *X3*, *Y3,* T_3 , in .3

Excluding T_2 and T_3 from equations (2.12)-(2.13), we find $\varepsilon_3 = \varepsilon_2 e^{-f_3(\theta_2+\theta_3)}$ (2-14)
At the CE section (Fig. 2.1., Fig. 2.6.) the thread has a horizontal shape and therefore, to determine the motion parameters of this region, we have the following equations

$X_4' = a\varepsilon_4$ (2.15) $T_4 = \rho_0 a^2 \varepsilon_4$ (2.16) $T_3 = T_4 e^{f_4\theta_3}$ (2.17)

Excluding T_3 and T_4 from equations (2.16)-(2.17), we find $\varepsilon_4 = \varepsilon_3 e^{-f_4\theta_3}$ (2.18)

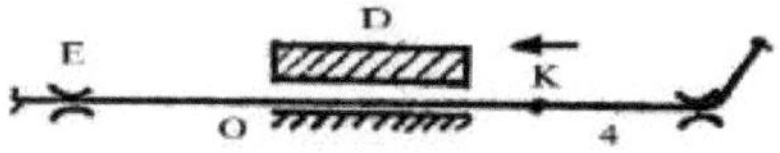

Figure 2.6. Schematic of the longitudinal wave motion at the CE section

The EF section of the filament forms the angle θ_5 (known in practice). The parameters of the area 5 of the filament (Fig. 2.1., Fig. 2.7.) are determined from the following equations

$X_5' = a\varepsilon_5 \cos\theta_5$ (2.19) $Y_5' = a\varepsilon_5 \sin\theta_5$ (2.20) $T_5 = \rho_0 a^2 \varepsilon_5$ (2.21) $T_4 = T_5 e^{f_5\theta_5}$ (2.22)

From these equations we find $\varepsilon_5 = \varepsilon_4 e^{-f_5\theta_5}$ (2.23)

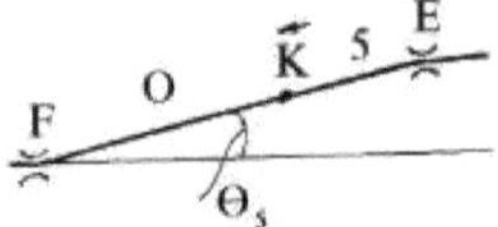

Fig. 2.7. Schematic of the longitudinal wave motion on the EF section

At section FN - the point where the yarn comes off the surface of the pack, the yarn initially has some curvilinear shape and is stress free.

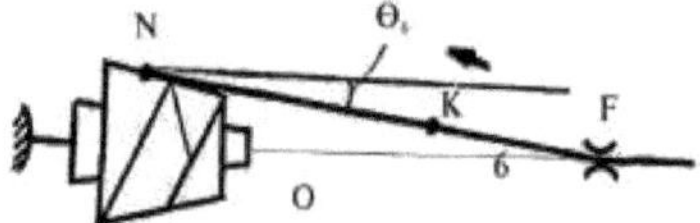

Fig. 2.8. Schematic of the longitudinal wave motion at the FN section

Let initially on the section FN the thread had a rectilinear shape (Fig. 2.1., Fig. 2.8.). Then, to determine the motion parameters of the area 6 of the filament, we have the following equations

$X_6' = a\varepsilon_6 \cos\theta_6$ (2.26) $Y_6' = a\varepsilon_6 \sin\theta_6$ (2.27)

$T_6 = \rho_0 a^2 \varepsilon_0$ (2.28) $T_5 = T_6 e^{f_6\theta_6}$ (2.29) $\varepsilon_6 = \varepsilon_5 e^{-f_6\theta_6}$ (2.30)

The assumption of straightness of the filament can take place only up to the moment of arrival of the longitudinal wave K at the point of filament convergence from the surface of the packing. At the vanishing point, the filament has two components of velocity tangential to the filament itself at the vanishing point and a circumferential velocity. And therefore, the shape of the filament motion in the FN region after the reflection of the longitudinal wave will not be rectilinear. As follows from equations (2.3, 2. 4),(2.7, 2.8, 2.9), (2.13, 2. 14), (2.17, 2.18), (2.22, 2.23), (2.29, 2.30) the tension and deformation of the filament reaches the maximum value in region 1, hence in the above formulation the breakage of the filament can occur only in region 1 at the moment of impact. The obtained solutions (2.3, 2.4), (2.7, 2.8, 2.9), (2.13, 2.14), (2.17, 2.18), (2.22, 2.23), (2.29, 2.30) should be considered as a scheme for calculating the motion parameters during the initial passage of the load wave. As follows from the analysis of the solutions obtained above and the results of numerical calculations, the highest tension is reached in region 1 and, therefore, thread breaks should occur immediately after the battle and only in region 1 - if the tension exceeds the permissible value at the moment of the grommet start. However, in practice, thread breaks occur in any of the regions 1 - 6 (Fig. 2.4.-2.9.). The main reasons for this may be the influence of reflection processes and interaction of longitudinal waves on the thread sections, which may result in: instantaneous tension increase from T to 2T, the influence of brake and compensator shocks, as well as boundary conditions and physical and mechanical properties of the thread itself. Below we consider the shock effect of the brake on the thread tension. Let at some moment of time t, after the wave

K moves to the left of region 4 (Fig. 2.1., Fig. 2.9.), the braking device is triggered. This formulation of the problem is quite reasonable, since the longitudinal wave K moves along the thread with a velocity of about 800 m/s. As a result of the impact of the braking device, two longitudinal waves P and Q appear in the thread (Fig. 2.9.). As a result, regions 7 and 8 appear.

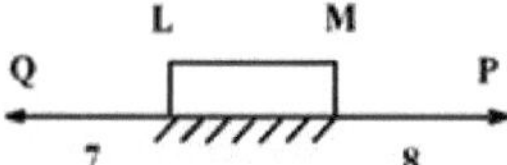

Fig. 2.9. Schematic of longitudinal wave motion at impact on a thread

Consider region 7. The longitudinal wave Q is an unloading wave. Indeed, the brake at the point L, there is a stop of motion of the thread particles. At the front of Q we have $X_7' - X_4' = a(\varepsilon_7 - \varepsilon_4)$ (2.31). But since $X_7' = 0,\ X_4' = a\varepsilon_4,$ we find 0 - as_4 = as_7 - as from equation (31)$_4$

As we can see, in region 7 the deformation $\varepsilon_7 = 0,$ hence $T_7 = \rho_0 a^2 \varepsilon_7 = 0,$ i.e. region 7 turns out to be a region of rest. Let us consider region 8. We prove that in region 8 the deformation is equal to twice the deformation of region 4. At the front P we have the following equation $X_8' - X_4' = a(\varepsilon_4 - \varepsilon_8)$ (2.32)

Obviously $X_8' = 0$ since the whole region 8 is the region of constant parameters and at the point M $X_8' = 0$. Substituting $X_8' = 0$ into equation (2.32) we obtain $0 - a\varepsilon_4 = a\varepsilon_4 - a\varepsilon_8$ Hence $\varepsilon_8 = 2\varepsilon_4$ (2.33)

Thus, at the moment of braking device impact in the right part 84 doubles and this becomes the cause of thread breakage in cases when 84>[e] or T8>[T]. Tables 2.1. and 2.2. show the results of calculations of tension and deformation of weft yarns in the interval of weft laying speeds from 20 to 26 m/s and coefficients of friction of yarns against guides at f=0.1 and f=0.2.

Table 2.1.

Results of calculations of weft yarns deformation by zones

№		Thickness deformation by zones in %					
	Laying speed, m/s.	£1	£2	£3	£4	£5	£6
1	20	2,5	2,47	2,41	2,38	2,35	2,32
			2,44	2,32	2,27	2,21	2,16
2	21	2,6	2,57	2,51	2,48	2,45	2,42
			2,54	2,42	2,36	2,30	2,25
3	22	2,8	2,76	2,70	2,67	2,64	2,61
			2,73	2,60	2,54	2,48	2,41
4	23	2,9	2,86	2,80	2,76	2,73	2,70
			2,83	2,70	2,63	2,57	2,51
5	24	h,o	2,96	2,89	2,86	2,82	2,79
			2,93	2,79	2,72	2,65	2,59
6	25	3,1	3,06	2,99	2,95	2,92	2,88

			3,03	2,88	2,81	2,74	2,68
7	26	h,h	3,26	3,18	3,14	h,n	3,07
			3,22	3,07	2,99	2,92	2,85

where the numerator is the coefficient of friction of the thread against the guides f = 0.1; the denominator is the coefficient of friction of the thread against the guides f = 0.2.

Table 2.2.

Results of weft yarn tension calculations by zone

№	Laying speed, m/s.	Tension of the weft by zones in cN.					
		T1	T2	Tz	T4	T5	T6
1	20	20,0	19,8	19,3	19,1	18,8	18,6
			19,5	18,6	18,2	17,7	17,3
2	21	20,8	20,6	20,1	19,8	19,6	19,4
			20,3	19,4	18,9	18,4	18,0
3	22	22,4	22,1	21,6	21,4	21,1	20,9
			21,8	20,8	20,3	19,8	19,3
4	23	23,2	22,9	22,4	22,1	21,8	21,6
			22,6	21,6	21,0	20,6	20,1
5	24	24,0	23,7	23,1	22,9	22,7	22,3
			23,4	22,3	21,8	21,2	20,7
6	25	24,8	24,5	23,9	23,6	23,4	23,0
			24,2	23,0	22,5	21,9	21,4
7	26	26,4	26,1	25,4	25,1	24,9	24,6
			25,8	24,6	23,9	23,4	22,8

where the numerator is the coefficient of friction of the thread against the guides f = 0.1; the denominator is the coefficient of friction of the thread against the guides f = 0.2.

The analysis of Tables 2.1 and 2.2 shows that the values of tension and deformations increase at the bobbin - weft section when the speed of weft laying increases and when the coefficient of friction of the yarn against the guiding working parts of the mechanism decreases. As mentioned above, the weft brake has a number of design drawbacks. Shock impact foot weft brake and the periodic nature of its movement contributes to cyclical changes in the weft thread, resulting in reduced reliability of the process of laying weft thread in the shed. In addition, the braking mechanism is a complex system. In this paper, the tension of the thread after the guide device is determined, which depends on the initial tension, the angle of coverage, the kinematics of the thread, the coefficient of friction of the guide (eye), but also on the geometric properties of the guide curve from the curvature of the curve. If the coverage angle of the thread guide is gradually increased, the thread tension naturally increases, with a constant initial thread tension. On the basis of increasing the angle of thread coverage of the guide, the shockless weft brakes are selected and developed, where in the period of acceleration and the beginning of sinker movement the thread friction force will be equal to zero, as the compensator 1 is in the lower position I, and there is no contact of the braking surface of the shoe 2 with the sinker 3 (Fig. 2.14) or no contact of the sinker 4 with the ball 3 (Fig. 2.15). Towards the end of sinker insertion, the compensator moving upwards (to position II, Fig. 2.14) brings the thread 3 into contact with the braking surface of shoe 2, and there is a gradual increase in the friction angle of the thread against the braking surface of the system. Consequently, the total tension of the thread T is made up of the constant minimum tension T, created by the foot and plate, and additional tension created by the friction material of the bracket and the position of the compensator.

л

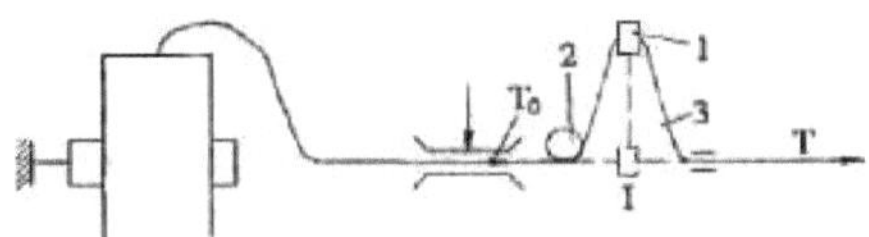

Fig. 2.14. Schematic diagram of the modernised weft brake

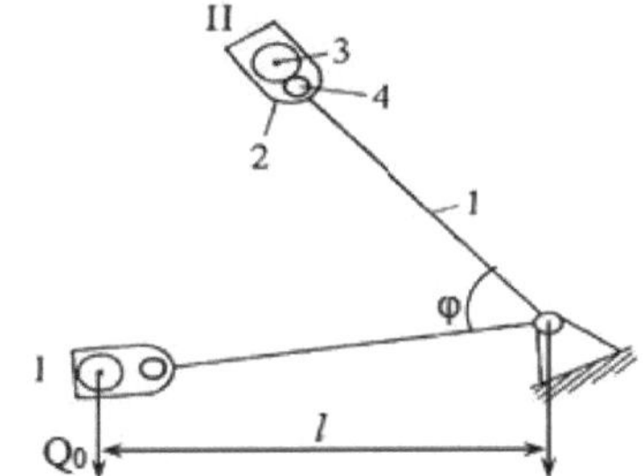

Fig. 2.15. Schematic diagram of modernised weft brake

During weft laying the compensator 1 takes the lower position I (Fig. 2.15), ball 3 moves away from thread 4, thread tension is minimal. At the end of weft laying, the compensator moves to position II, the ballZ gradually presses on the thread4. The maximum pressure of the ball on the thread is in the topmost position II, therefore, the thread tension will be maximum during this period. The analysis of oscillograms (Fig. 2.16.) shows that the maximum peak is obtained at the moment of the beginning of winding the weft thread from the stationary bobbin, but this peak is 1.5-2 times less than in the existing design (see Fig. 2.2.). Let us characterise by zones the tension of the weft yarn during the laying process with the modernised design of the weft brake (Fig. 2.16.). Zone 1 starts at the moment of weft weaver acceleration (140° of the main shaft). The zone ends when the weft weaver reaches its maximum speed. Together with the weft weaver, a certain section of the yarn is also moving, the speed changes from zero to 20-26 m/s in a very short time (0.006 sec). The tension on this section changes from zero to some value, not significant on the oscillogram.

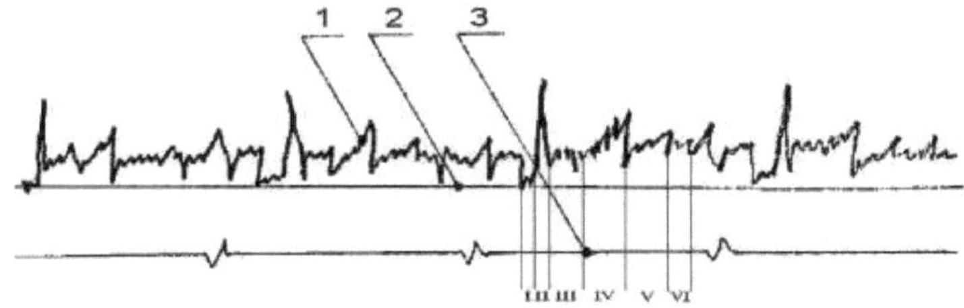

Fig. 2.16. Oscillogram of the weft thread tension

The thread following the grommet is selected from the free-flying section prepared by the compensator. Zone 2 corresponds to the beginning of the flight of the weaver with the weft thread following it through the shed. It starts at the moment of decreasing yarn tension, when the compensator has not yet selected the yarn reserve. Zone 2 ends before the first maximum peak in the tension curve. This tension is the maximum tension, which is created by the inertia of the thread and the brake washers. The driving yarn changes to 26 m/s in a very short time, overcoming the resistance of the brake washers, for the process of inserting the weft into the shed allows the peak tension to be reduced by a factor of about 1.5 to 2 times that of the existing brake. The tension in this zone depends on the advance angle of the compensator, the linear density of the weft yarn, the yarn weight, the section with slackness, the homogeneity of the yarn slackness and the initial tension of the brake washers. Zone 3 is characterised by the fact that the tension stabilises after a slight decrease and is kept more uniform throughout the interval. The tension in this area depends on the conditions of winding from the feed bobbin, the coefficient of friction of the yarn on the guide elements and the angles at which the yarn covers them. This period lasts until the compensator driver starts to lift. Zone 4 starts with a certain increase in yarn tension, the value of which depends on the linear density of the weft, the speed of the weft weaver, the conditions of winding from the feeding bobbin, and the spring pressure on the brake washer. The tension in this zone is the sum of the tension in zone 3 plus the braking effect of the brake angle profile. Zone 5 of the weft shedding cycle through the shed begins with the moment when the weaver flies under the first brake of the pick-up box. This moment is characterised by a sharp decrease of the laying speed and the initial section of the moving yarn. On the thread tension curve, this moment corresponds to a decrease in the tension of the brake angle with the thread, with further lifting of the compensator 9, the tension gradually increases to a certain value. The yarn tension in this zone depends on the tension in the previous zones, the speed of arrival of the weeder, the linear yarn density, the spring pressure on the washers and the geometrical shape of the brake angle profile. Zone 6 of the tension curve starts with the beginning of the weaver return. It is characterised by a constant tension over the entire interval with very slight fluctuations at certain points. Zone 6 is particularly necessary to create a certain tension in the weft yarn. After the weft spacer returns, the centred weft yarn, which has a certain tension required for normal fabric formation, is gripped by the jaws of the weft spacer. The section of thread, which is in the zone and is held by the grippers of the selvedge-former, is cut off from another section winding off the bobbin, the end of the weft thread is withdrawn by the weft returner to the initial position for transferring to the weaver. The method of calculation of changes in the parameters of yarn movement in interaction with the existing mechanisms of weft laying and braking are considered in the section, which is based on the theory of wave propagation in an elastic medium. The modernised design of the weft braking mechanism excludes shock effects on the weft yarn in the CE section (see Fig. 2.6) and provides for the installation of the brake shoe 2 (Fig. 2.14) at point C in the CE section (see Fig. 2.6) Otherwise, the calculation is similar. Let us consider the influence of the modernisation in the filling on the tension and deformation of the threads on the CE section (Fig. 2.17).

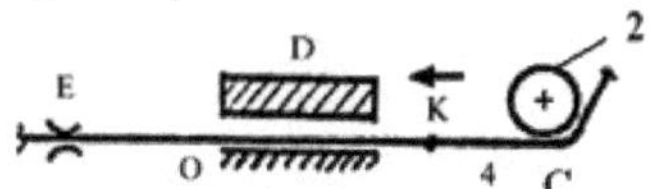

Figure 2.17.

On the CE section, to determine the motion parameters of this region, we have the following equations $\dot{X} = a\varepsilon_4$ (2.15`) $T_4 = \rho_0 a^2 \varepsilon_4$ (2.16`)

$T_3 = T_4 e^{f_4 \theta_3}$ (2.17`), where f) is the coefficient of friction of the thread against brake shoe 2; θ_3 is the angle of friction of the thread against brake shoe 2. Excluding T_3 and Td from equations (2.16') and (2.17'), we determine $\varepsilon_4 = \varepsilon_3 e^{-f_4 \theta_3}$. Tables 2.3 -2.4 show calculations of deformation f and tension Td of the modernised braking system when changing the friction angle 0_3 and the coefficient of friction of the thread about the working bodies of the mechanism. The analysis shows that as the friction angle of the filament against the brake shoe increases, the deformation and tension of the sinker in the section in front of the modernised system decreases. This reduction is most pronounced as the friction coefficient of the yarn against the braking body of the mechanism increases. As the weft threading speed increases, the warp and tension of the weft increases.

Table 2.3.

V = 20m/sec

№	Value of the coefficient of friction of the thread against the brake shoe, f	Deformation ed and tension Td of the sink at friction angle 0z degrees				
		14	28	42	56	70
1	0,1	2,35	2,30	2,24	2,18	2,13
		18,8	18,4	17,9	17,4	17,0
2	0,2	2,30	2,19	2,08	1,98	1,89
		18,4	17,5	16,6	15,8	15,1
3	0,3	2,24	2,08	1,93	1,80	1,67
		17,9	16,6	15,4	14,4	13,4

4	0,4	2,18	1,98	1,80	1,63	1,48
		17,4	15,8	14,4	13,1	11,8
5	0,5	2,13	1,90	1,67	1,48	1,31
		17,1	15,8	13,4	11,8	10,5
6	0,6	2,08	1,80	1,55	1,34	1,16
		16,8	14,4	12,4	10,7	9,3

where numerator is the value of deformation unit denominator is the value of tension Td

Table 2.4.

V = 23m/sec

№	Value of the coefficient of friction of the thread against the brake shoe, f	Deformation ed and tension Td of the sink at friction angle 0z degrees				
		14	28	42	56	70
1	0,1	2,73	2,67	2,60	2,54	2,48
		21,8	21,4	20,8	20,3	19,8
2	0,2	2,67	2,54	2,42	2,30	2,19
		21,4	20,4	19,4	18,4	17,5
3	oh,h	2,60	2,42	2,25	2,09	1,94
		20,8	19,4	18,0	16,7	15,5
4	0,4	2,54	2,30	2,09	1,89	1,72
		20,3	18,4	16,7	15,1	13,8
5	0,5	2,48	2,19	1,94	1,72	1,52
		19,8	17,5	15,5	13,8	12,2
6	0,6	2,42	2,09	1,80	1,56	1,35
		19,4	16,7	14,4	12,5	10,8

where numerator is the value of deformation unit denominator is the value of tension Td

Table 2.5.

V = 26m/sec

№	Value of the coefficient of friction of the thread against the brake shoe, f	Deformation £4 and tension T4 of the sink at friction angle 0z degrees				
		14	28	42	56	70
1	0,1	3,10	3,03	2,96	2,88	2,81
		24,8	24,2	23,7	23,0	22,5
2	0,2	3,03	2,88	2,75	2,62	2,49
		24,2	23,0	22,0	21,0	19,9
3	oh,h	2,96	2,75	2,55	2,37	2,21
		23,7	22,0	20,4	19,0	17,7
4	0,4	2,88	2,62	2,37	2,15	1,95
		23,0	21,0	19,0	17,2	15,6
5	0,5	2,81	2,49	2,20	1,95	1,73
		22,5	19,9	17,6	15,6	13,8
6	0,6	2,75	2,37	2,05	1,77	1,53
		22,0	19,0	16,4	14,2	12,2

where numerator is the value of deformation unit

the denominator is the value of tension T.(

The weft tension must be different at certain periods of the weaving machine. The weft tension should be minimised at the beginning of the weft insertion and at the end of the weft insertion there should be additional tension for weft braking, which prevents loops in the shed on the take-up box side. When the weft weaver is returning, the weft weft should be tightened with the compensator, while the weft weft has maximum braking to prevent the weft from looping and winding off the bobbin. The tension of the weft weft weft yarns before the weft take-up can be determined by the following expression $T = T_o + T_K + T_T$ (2.34)

where T_o - weft pre-loading tension created by the permanent braking device; T_K - weft tension depending on weft yarn stiffness, friction coefficient and angle of friction in the compensator eye; T_T - additional weft tension created by the rolling body (load) or friction of the braking surface of the shoe and the position of the compensator. The preliminary filling tension is determined by

$T_o = Nf_1 + Nf_2 = N(f_1 + f_2) = 2Nf$ (2.35)

where N - normal pressure on the thread in the braking device; f, f - friction coefficients of the thread against the upper and lower plate, respectively. According to L Euler, the relation between the incoming branch T_o and the escaping branch T_K has the following expression and depends on the friction angle (a) and the friction coefficient (/) of the filament against the compensator eye (Fig. 2.18). $T_K = T_o$ - ehr $(f \cdot \alpha)$ (2.36). . It follows from Fig. 2.18 that the angle of friction can be determined as follows

$$\alpha = \varphi + \beta,$$

where $\varphi = (90^0 - \varphi_1)$, $\beta = (90^0 - \beta_1)$, $tg\varphi_1 = \frac{l_1}{S}$, $tg\varphi_2 = \frac{l_2}{S}$

Considering that '/ and l_2 are constant values and can be determined practically on the weaving machine $li = 1_2^{=}$ 75 mm and that the compensator movement S has a range from O to 200 mm, it is possible to determine the values of the angles φ, φ_1, β, β_1 and the friction angle a of the weft against the compensator eye.

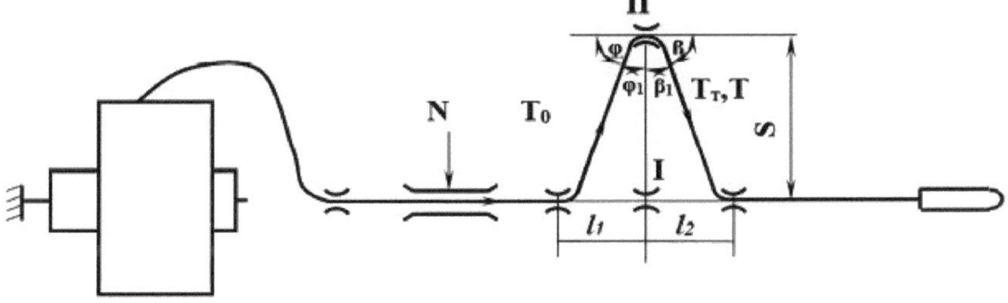

Fig. 2.18. Calculation of weft yarn tension before the weft bobbins

Table 2.6 shows the results of calculating the angles φ, φ_1, β, β_1 and the yarn friction angle in the eye depending on the position of the weft compensator.

Table2.6.

Angle calculation results depending on the compensator position

№	Angles in deg	Compensator displacement, S, mm										
		0	20	40	60	80	100	120	140	160	180	200
1	Φ1	90	75	62	51	43	37	32	28	25	23	21
2	Φ	0	15	28	39	47	53	58	62	65	67	69
3	Ᵽ1	90	75	62	51	43	37	32	28	25	23	21
4	Ᵽ	0	15	28	39	47	53	58	62	65	67	69
5	a	0	30	56	78	94	106	116	124	130	134	138

It follows from Table 2.6 that the maximum friction angle a of the yarn against the eye at the topmost position of the weft compensator. Euler's formula (2.36) gives the same thread tension $T_К$ for a given girth angle a and tension of the incoming branch $T_о$ regardless of the shape of the guide eye on which the weft compensator is located. For example, for circular cylinders with different diameters and at the same girth angles, the thread tension is the same. What is certain is that the thread tension cannot be the same for different shapes of the cylinder guide on which the sinker is located. It may be greater for some cylinder shapes and less for others. Consider the case where the cylinder guides are circular in shape. For a thread of length *l* sliding along the circle, with the arc of coverage equal to (g-a*)* and depending on the stiffness of the weft, the thread tension ($T_К$) is as follows

type

$$T_к = \frac{2K_н \cdot r \cdot f}{1+f^2}\left(e^{f\alpha} + \frac{1-f^2}{2f} \cdot \sin\alpha - \cos\alpha\right) \qquad (2.37)$$

where K_H is the stiffness of the weft yarn, depending on the fibre type and yarn linear density, cn/mm; *f is* the coefficient of friction of the yarn against the compensator guide eye; a is the angle of friction of the yarn against the compensator guide eye; d is the radius of friction of the yarn against the compensator guide eye.

According to the formula (2.37) we have calculated the tension of the weft $T_К$ depending on the compensator position at different values of friction radius, *friction* coefficient and stiffness of weft yarns (Appendix 1), which are presented in Fig.2.19-2.22. From the graphs it follows that in all variants with increasing parameters g, *f* $K_Н$ the tension of the weft increases, with the greatest influence of friction coefficient and stiffness of yarns. The thread tension $T_Т$, created by the rolling element of the compensator according to Fig. 2.15 is determined by $T_T = Q \cdot (f_1 + f_2)$ or $T_T = 2 \cdot Q \cdot f$ (2.38)., where Q is the normal pressure on the rolling element sinker.

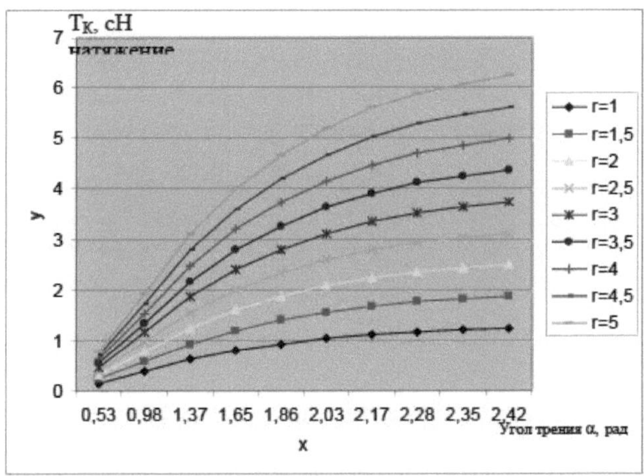

Fig. 2.19. Graph of change of the sinking tension as a function of the friction angle at different friction radii

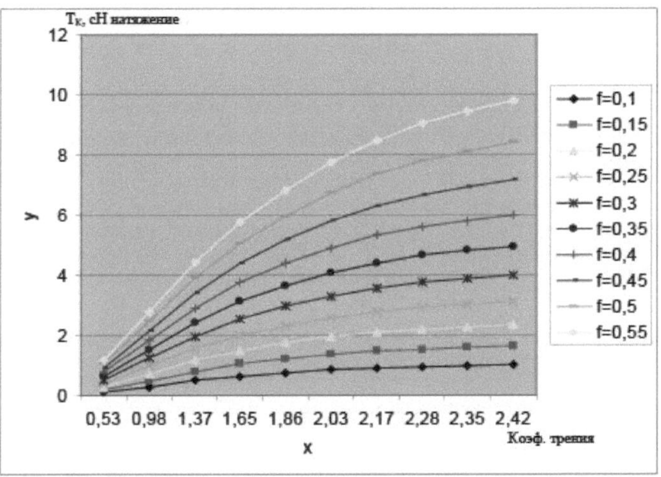

Fig. 2.20. Graph of change in the tension of the sinker depending on the angle of different friction coefficients

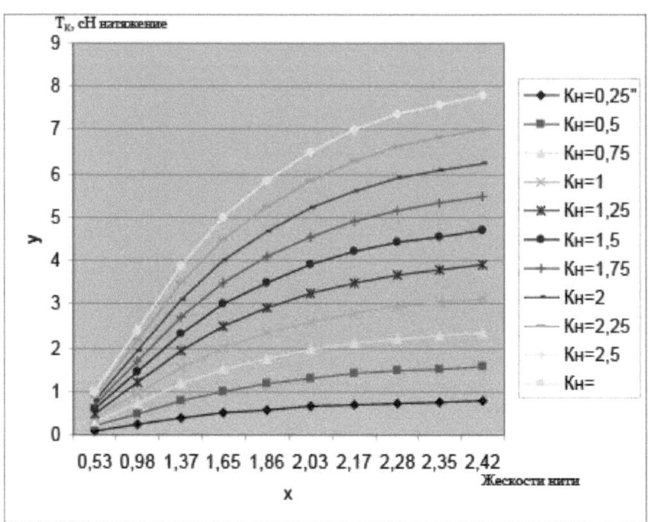

Fig. 2.21.Graph of change of the weft tension as a function of the friction angle
at different thread stiffnesses

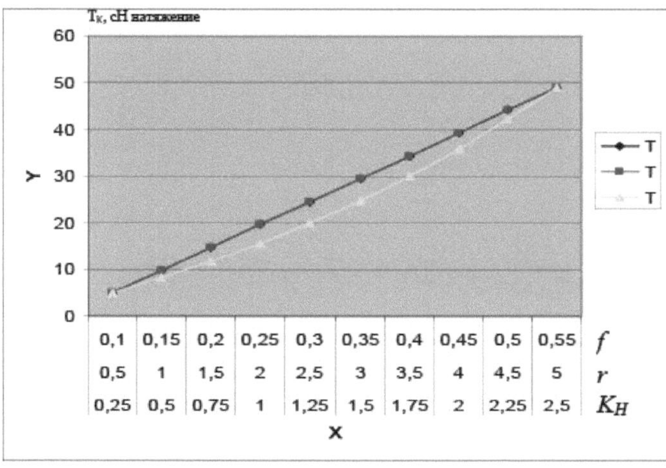

Fig. 2.22. Graph of change in the tension of the weft as a function of the friction angle at different friction radii, friction coefficient, thread stiffness

In this equation, the normal pressure has a variable value and depends on the position of the compensator, i.e. the swing angle y (Fig. 2.15)

$$Q = Q_0 \cdot \sin\psi = Q_o \cdot \frac{S}{l} \quad (2.39) \qquad \sin\psi = \frac{S}{l}$$

Substituting (2.39) into (2.38) we obtain

$$T_T = 2Q_o \cdot \frac{S}{l} \quad (2.40)$$

The compensator displacement value is known S=0-200 mm., Length of the compensator l = 203 mm. Table 2.7 shows the calculations of the swing angle of the compensator (ψ) and the normal pressure of the load on the thread (Q).

Table 2.7.

Calculation results of the compensator swing angle and variable load pressure on the string

№	Name		Compensator displacement, mm, S									
			20	40	60	80	100	120	140	160	180	200
1	Swing angle of the compensator y deg		5	11	17	23	29	36	43	52	62	80
2	arcsin y		0,098	0,197	0,296	0,394	0,493	0,591	0,69	0,788	0,887	0,985
3	Weight of the load gr.	5	0,49	0,99	1,48	1,97	2,46	2,96	3,45	3,94	4,44	4,93
4		10	0,98	1,97	2,96	3,94	4,93	5,91	6,90	7,88	8,87	9,85
5		15	1,47	2,96	4,44	5,91	7,40	8,87	10,35	11,82	13,31	14,78
6		20	1,96	3,94	5,92	7,88	9,86	11,82	13,80	15,76	17,74	19,7

Substituting (2.40), (2.37) and (2.35) into (2.34) we obtain the total tension of the weft yarns during the weaving process.

$$T = 2Nf + \frac{2K_{н} \cdot r \cdot f}{1+f^2}(e^{f\alpha} + \frac{1-f^2}{2f} \cdot \sin\alpha - \cos\alpha) + 2Q_o \cdot f \frac{S}{l} \qquad (2.41)$$

As can be seen, formula (2.41) takes into account the stiffness of the weft (type and linear density of the thread), friction coefficient, friction radius, position of the compensator and pressure of the weight on the thread in the compensator. Table 2.8 shows the results of calculation of the filling tension (T_o), tension due to the stiffness of the thread ($T_к$) at the compensator eye, tension created by the weight ($T_т$) in the compensator, with the following parameter values *f=0*,25; **=10 сн; Q₀=10 сн,** $K_н = 1\frac{сн}{мм}$; values a are taken from Table 2.7. *mm.*

Table 2.8.

Calculation results of sinker tension depending on the compensator position

Weft yarn tension, cf.	Compensator displacement, mm, S									
	20	40	60	80	100	120	140	160	180	200
Filling tension of the sink, CH., Tho.	5	5	5	5	5	5	5	5	5	5
Utochina tension from yarn stiffness in the compensator cn., $T_{.к}$	1,4	2,7	h,h	4,0	4,3	4,4	4,5	4,5	4,5	4,5
Tensioning of the weir i compensator From load, cf., $T_{.т}$	0,5	1,0	1,5	2,0	2,5	h,o	3,5	4,0	4,5	4,9
Total tension of the utocina, cn.,T.	6,9	8,7	9,8	n,o	П,8	12,4	13,0	13,5	14,0	14,4

Analysis of the table shows that the total tension (T) of the weft increases as the compensator is moved upwards. A device for measuring the coefficient of friction of textile materials is proposed, which has a dynamometer located on a movable platform and connected to the end of a textile thread, and the textile thread is in contact with the material under study, stacked in the longitudinal groove of the cylinder. The other end of the textile thread is loaded with a constant load. The material under study, depending on the friction pairs, can be made of metal, non-metal (plastic, wood, rubber, etc.), textile fibres (cotton, wool, linen, silk and their mixtures, etc.). Fig.2.17 shows a device for measuring the coefficient of friction of textile materials.

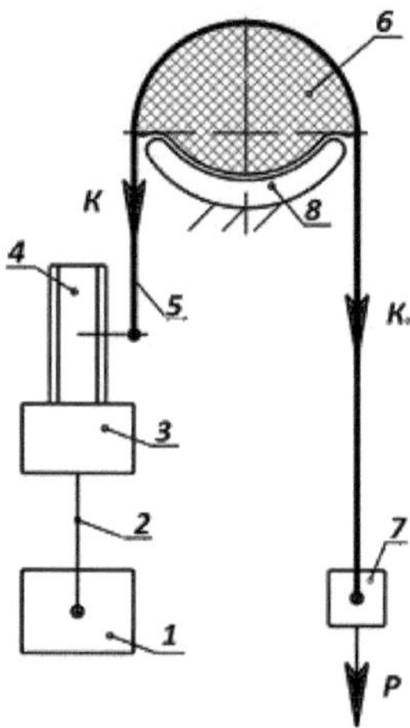

Fig. 2.17. Schematic diagram of the device for measuring the coefficient of friction of textile materials.

The device contains a drive 1, a screw connection 2, a movable platform 3 and a dynamometer 4.A textile thread 5 is thrown over the surface of the material under study 6, with one end (the leading branch) connected to the dynamometer 4 and the other end (the slave branch) to a constant weight 7. The material under test 6 is placed in a longitudinal groove of the stationary cylinder 8. The drive 1 moves the platform 3 downwards by means of the screw connection 2, and together with it the dynamometer 4. Textile thread 5 is on the surface of the material 6 in a state of equilibrium due to the constant weight 7. Further lowering of the dynamometer 4 leads to sliding of the thread 5 on the surface of the material 6. At the moment of touching (the beginning of sliding) of the textile thread 5 on the surface of the investigated material 6 we take readings from the scale of the dynamometer 4, which corresponds to the tension K *of* the leading branch of the thread 5, and the value of the constant weight 7 corresponds to the tension K_o of the leading branch of the thread 5.

The coefficient of *frictionf* on the contact surface between the textile thread 5 and the material under test is calculated according to the formula at $\alpha = \pi$

$$f=\frac{\lg K-\lg K_0}{\alpha\cdot\lg e} \qquad (2.1)$$

where: K - tension of the leading branch of the thread 5 is determined by the dynamometer scale 4; $K_э$ - tension of the slave branch of the thread 5 is determined by the weight of the constant weight 7; a - angle of girth (friction angle) of the material 6 by the textile thread 5, in radians.

The moment of touching (beginning of sliding) of the textile thread 5 on the surface of the test material 6 corresponds to the friction coefficient at rest. Further uniform sliding of the textile thread 5 on the surface of the test material 6 corresponds to the coefficient of friction in motion. An abrupt change of speed (by means of drive 1) of movement of the textile thread 5 on the surface of the investigated material 6 corresponds to the coefficient of friction from the sliding speed. Similarly determine the coefficient of friction at rest, in motion and sliding speed for the rubbing friction pairs of the thread 5 and the material 6 of different structures, by fitting the latter into the grooves of the stationary cylinder 8. This is due to the fact that the thread in motion contact with the working bodies of machines, devices and appliances with different speeds, loads and friction bodies (metal, plastic, wood, rubber, etc.). In addition, in products thread and fibre interact with each other and contact with textile materials (cotton, wool, linen, silk and their mixtures, etc.) in the formation of knitted, woven and non-woven fabrics. The dimensions of the cylinder 8 are determined by the dimensions of the material 6 being examined, which should not exceed 2/3 of the diameter of the material 6. Since exceeding this size leads to additional friction of the textile thread 5 against the cylinder 8, which leads to a significant distortion of the results of the study of the coefficient of friction. Repeatability of the research is provided by reversibility of the drive 1, i.e. installation of the platform 3 and dynamometer 4 in the initial (upper) position through the screw connection 2. The value of the friction coefficient was calculated on the basis of formula (2.1). The experiments were carried out at a constant friction angle equal to $18O^0$, i.e. a = 1. Table 2.8 shows the results of calculation of friction coefficient for different diameters and friction surfaces Co=1Ogr and $\alpha = \pi.$, different linear densities, cotton threads and natural silk threads. It follows from Table 2.8 that at the beginning of sliding of the thread on the surface, corresponding to the coefficient of friction at rest in all variants of the experiment is greater than the values of the coefficient of friction in motion, corresponding to the state of the friction pair (thread-surface) uniform motion. The smallest value of the friction coefficient when using metal on the friction surface, and the largest when using rubber on the friction surface. As the friction radius of the friction surface and the linear density of the yarns increases, the coefficient of friction will increase. Natural silk threads have a lower coefficient of friction than cotton threads.

Table 2.8.

Results of the calculation of the friction coefficient depending on the diameter and condition of the rubbing surfaces and yarn line density

№	Linear yarn density, tex yarn type	Condition of the friction pair	Diameters of rubbing surfaces, mm					
			2	3	4	5	6	8
1	15.4 cotton	At the beginning of the movement, at rest	0,22 0,41	0,25 0,44	0,29 0,47	0,32 0,51	0,35 0,54	0,37 0,55
		Uniform Motion	0,20 0,39	0,22 0,41	0,25 0,44	0,28 0,46	0,32 0,51	0,35 0,54
2	15x2 cotton	At the beginning of the movement, at rest	0,28 0,48	0,29 0,48	0,32 0,51	0,34 0,53	0,37 0,55	0,39 0,58
		Uniform Motion	0,27 0,44	0,28 0,46	0,29 0,48	0,34 0,49	0,34 0,53	0,36 0,56
3	15x3 x/b	At the beginning of the movement, at rest	0,32 0,51	0,34 0,52	0,36 0,55	0,38 0,57	0,40 0,59	0,43 0,62
		Uniform Motion	0,29 0,48	0,32 0,51	0,35 0,54	0,36 0,55	0,38 0,57	0,40 0,59
4	15x4 x/b	At the beginning of the movement, at rest	0,37 0,55	0,38 0,57	0,40 0,59	0,41 0,59	0,43 0,61	0,45 0,64
		Uniform Motion	0,35 0,54	0,37 0,55	0,38 0,57	0,38 0,57	0,40 0,58	0,41 0,60
5	2,33x5 n/w	At the beginning of the movement, at rest	0,19 0,38	0,20 0,39	0,20 0,41	0,25 0,44	0,27 0,46	0,29 0,49
		Uniform Motion	0,15 0,33	0,17 0,35	0,19 0,37	0,22 0,40	0,24 0,43	0,28 0,47
6	2,33x10 n/w	At the beginning of the movement, at rest	0,25 0,44	0,27 0,45	0,29 0,48	0,30 0,48	0,32 0,51	0,35 0,55
		Uniform Motion	0,22 0,43	0,24 0,43	0,27 0,45	0,28 0,46	0,30 0,49	0,33 0,51
7	2,33x15 n/w	At the beginning of the movement, at rest	0,30 0,49	0,32 0,51	0,34 0,53	0,37 0,56	0,38 0,57	0,41 0,60
		Uniform Motion	0,28 0,47	0,29 0,48	0,32 0,51	0,33 0,51	0,35 0,54	0,38 0,58
8	11,6x2 n/a	At the beginning of the movement, at rest	0,25 0,43	0,27 0,45	0,28 0,46	0,30 0,49	0,33 0,51	0,36 0,55
		Uniform Motion	0,22 0,41	0,22 0,42	0,25 0,44	0,28 046	0,30 049	0,34 0,53

where: numerator - values of the coefficient of friction of thread against metal; denominator - values of the coefficient of friction of thread against rubber.

The third chapter contains "Investigations of warp tension on micro-spacer looms for shirt weaving". On the micro-spacer weaving machine with a shuttleless weaving machine, the warp tension is regulated during one machine cycle by means of a movable scalo system. More precisely, the increase of the warp tension during shed formation is compensated by the deflection of the scalo. However, the operating experience of these machines, as well as a number of studies have shown that the mechanism of the swinging scalp in serial machines does not fully compensate for warp deformation during shedding. Different patterns of repeated yarn loading on the loom during operation have different effects on the intensity of plastic elongation accumulation, i.e. yarn fatigue. The least fatigue of cotton yarns is caused by multiple loading, where in each cycle the maximum load is applied to the yarn for a short period of time, while most of the period is spent on resting the yarn in the unloaded state (Fig. 3.1). Consequently, in order to preserve technologically useful yarn properties, the weaving process should be constructed in such a way that the change in yarn tension is close to the curve in Fig. 3.1, i.e. the loading should be short in time and the rest in the unloaded state should be relatively long. In addition, it is necessary to strive to reduce the absolute magnitude of loads and to reduce the amplitude of their variation so that the product Pmax(Pmax-Pmin) is the smallest. The warp tension diagram in Fig. 3.2 shows that the warp is under a relatively high load most of the time during the weaving machine cycle, while the warp is under a minimum load only during a very short cycle time during the run-in period.

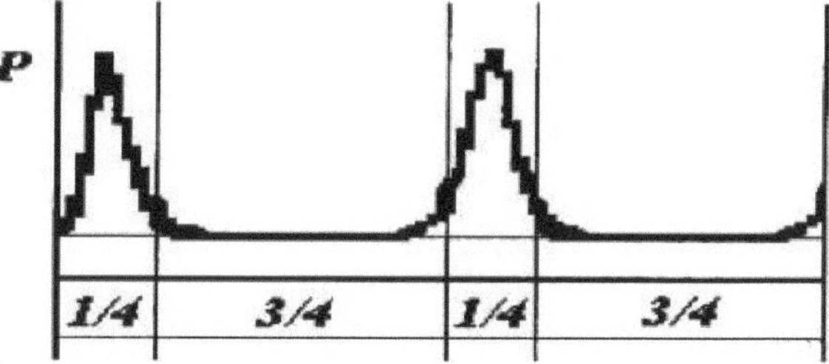

Fig. 3.1. Regularity of multiple loading on the yarn during weaving machine operation

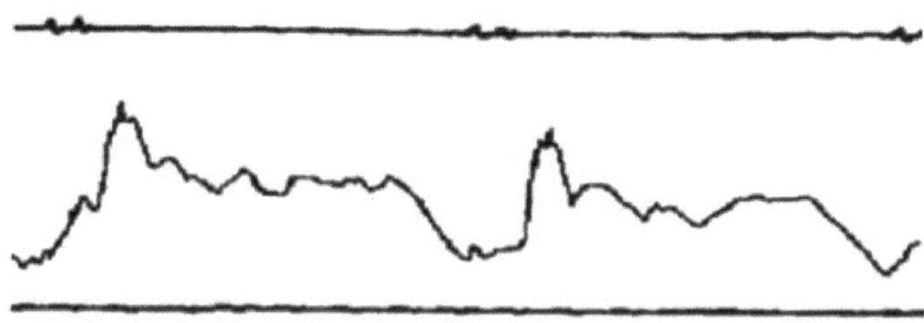

Fig.3.2: Oscillograms of warp tension for the existing scalo system

To reduce yarn fatigue, the shedding tension should be reduced as shown in the curve in Fig. 3.1. During shedding, a high warp tension is undesirable. It must be such that the open shed maintains a given shed height and that the weakly stretched yarns do not stick. One method of obtaining this law of warp tension variation is the method of forced rolling of the scalo. This method consists in reducing the total filling tension of the warp to the lowest possible value. However, at the moment of weft surfing to the fabric edge, the warp tension is forcibly increased to the level required to create normal weft surfing conditions. As a result of the work on optimising the warp tension change on the weaving machine, we have proposed a new design of a moving scalo system (Fig. 3.3), consisting of a scalo 1 located on the shoulders of three-shoulder levers 2 placed in the bracket bearings. The second arms of levers 2 are loaded by spring 3, and the third arm is kinematically connected with electromagnet 4 by means of rod 5 and armature 6. The operation of the electromagnet 4 is controlled by means of a synchroniser of the loom cycle, which is made in the form of a textolite disc 7, located on the main shaft and a permanent magnet 8 with a certain arc length embedded on the surface of the disc 7. The magnet 8 acts on the reed switch 9 of the synchroniser connected by means of controlling the electromagnet, made in the form of a controlled rectifier 10. This scalo system operates according to the method of forced rocking of the scalo. When producing the fabric, the filling tension of the warp threads is set at a minimum level, sufficient only to ensure that when the shed is open, the height of the shed is maintained, and at the moment of scoring there would be no sticking of weakly stretched threads. For this reason, during shedding, the increase in warp tension is insignificant and tends to zero. In the shedding phase, the disc 7 brings the permanent magnet 8 to the reed switch 9 and closes the contact. As a result, the machine cycle synchroniser supplies a pulse to the controlled rectifier 10, which supplies power to the coil of the electromagnet 4. The pole lugs attract the armature 6, which through the rod 5 and the additional arm of the lever 2 creates an additional moment on the mobile system of the scalo, which leads to an increase in the tension of warp threads to the value required for the surf of the weft to the fabric edge. After the surf, the disc 7 moves the permanent magnet 8 away from the reed switch 9, the contact of which opens and the rectifier de-energises the coil of the electromagnet. Consequently, in the phase of the weft surfing to the fabric edge, the moment created by the springs 3 and the additional moment created by the electromagnet 6 act on the mobile system of the scalo, while in the phases of scoring and shedding only the moment created by the springs 3 acts. When the warp thread breaks, the machine stops in the backstitch position. But in this case, only the moment created by the springs 3 acts on the moving system, as the coil of the electromagnet will be de-energised, although at this moment the permanent magnet acts on the reed switch. When changing the assortment of produced fabrics and changing the tension required for the weft to the fabric edge, the value of the additional torque created by the electromagnet 4 can be adjusted by changing the coil current or the arm length of the lever 2.

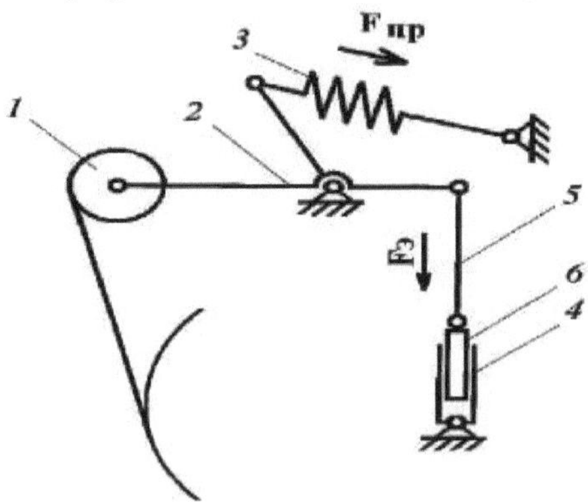

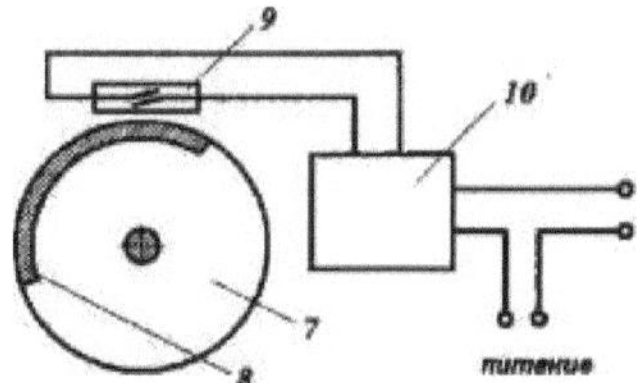

Fig.3.3 New design of the mobile scalo system.

As a result of experimental investigations of the new scalo system, oscillograms of warp tension changes on a weaving machine with microspacers filled with plain weave shirt fabric are obtained, shown in Fig. 3.2 - for the existing scalo system and Fig. 3.4 - for the new scalo system. 3.4 for the new scalo system.

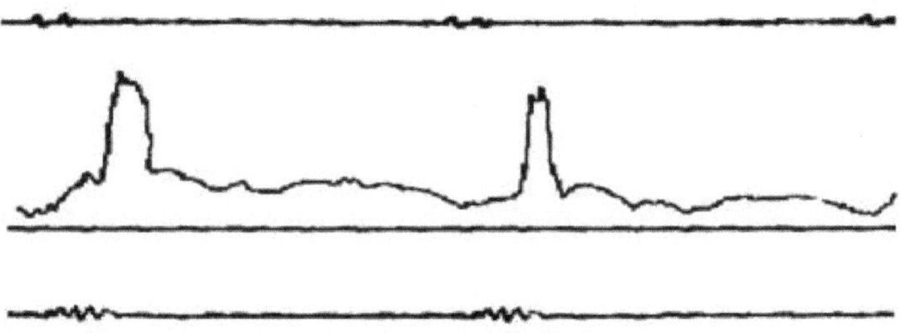

Fig.3.4: Warp tension oscillograms for the new scalo system The results of processing of experimental studies, the nature of warp tension variation per loom cycle are shown in Table 3.1.

Table 3.1.

Changes in warp tension per loom cycle

Name	Warp tension, cN		
	Zastazhd	Yawning	Surf
The existing rock climbing system	30,5	34,6	54,7
The new rock climbing	21,8	23,6	54,4

The analysis of the data shows that the weaving machine with the new scalo system was able to reduce the warp tension by 29 % at stalling and by 32 % at shedding, while maintaining the warp tension at surf. An analytical study of warp tension per loom cycle, i.e. per revolution of the main shaft of the weaving machine, is also presented. The differential equation of motion of the movable scalo system has the following form

$$J\varphi = M_k - M_F - M_M \qquad (3.1)$$

where *J is the* total moment of inertia of the links of the scalo moving system reduced (to the axis of rotation of the podokalina); φ - is the angular acceleration of the podokalina; M_k *is the* moment of the warp tension force relative to the axis of rotation of the podokalina; M_F is the moment of the spring elasticity force relative to the axis of rotation of the podokalina; M_M is the moment of the electromagnet force relative to the axis of rotation of the podokalina.

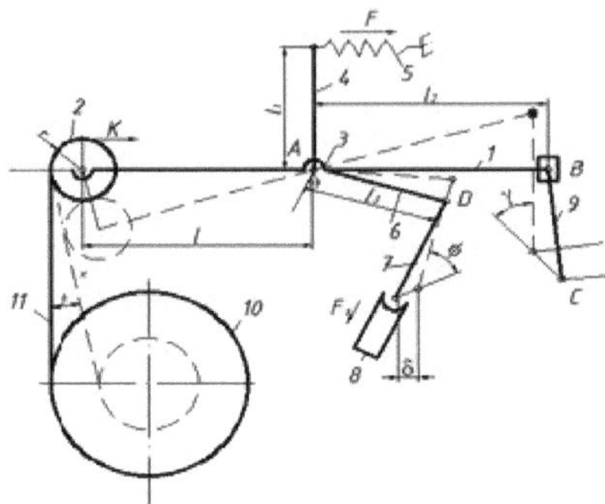

Figure 3.5: Calculation of warp tension.

From Figure 3.5 we find $M_k = K(l+r)\cos\beta - Kr = K[(l+r)\cos\beta - r]$, (3.2)

$M_F = 2Fl_1 = 2(F + C\varphi l_1)l_1 = 2Fl_1 + 2Cl_1^2\varphi$, (3.3) $F = C\lambda$; $M_M = F_M \cdot l_M$.

where: C - coefficient of spring stiffness, kG/cm; φl_1 – values of spring compression when rotating the movable system at the angle φ around the axis of the podokalina; K - tension of warp threads, kG; *li* - shoulder of spring action, cm; l - distance from the axis of the scalo to the axis of the podokalina, cm; l_M - shoulder of electromagnet action on the podokalina, cm; r - radius of the scalo, cm; F - force of tightening of the cargo spring, kg; F_3 - force of electromagnet action, kg.

Substituting the value of moments in equation (3.1), we obtain

$$J\varphi = K[(l+r)\cos\beta - r] - 2Fl_1 - 2Cl_1^2\varphi - F_M \cdot l_M \qquad (3.4)$$

Or

$$J\varphi + 2Cl_1^2\varphi = K[(l+r)\cos\beta - r] - 2Fl_1 - F_M \cdot l_M \qquad (3.5)$$

Dividing this equation by J we have

$$\varphi + \frac{2Cl_1^2\varphi}{J} = \frac{K[(l+r)\cos\beta - r] - 2Fl_1 - F_M \cdot l_M}{J}; \qquad (3.6)$$

Let's denote $P^2 = \frac{2Cl_1^2}{J}$ $\quad b = \frac{K[(l+r)\cos\beta - r] - 2Fl_1 - F_M \cdot l_M}{J}$;

Substituting these notations into (4), we obtain $\varphi + P^2 \cdot \varphi = b$

The general solution of this inhomogeneous differential equation with constant coefficients has the following form:

$$\varphi = A\cos Pt + B\sin Pt + \frac{b}{P^2}; \qquad (3.7)$$

where: A and B are integration constants.

Let us determine the constants A and B from the initial conditions. At the initial moment of time the mobile system is at rest. Therefore, the initial conditions are $\varphi = 0,\ \varphi = 0$ at $t = O$

Using the first initial condition = 0 at /=0

we obtain: $A + \frac{b}{P^2}$; $A = -\frac{b}{P^2}$ Using the second initial condition $\varphi = 0$ at t

we have $\varphi = -AP\sin Pt + B \cdot P\cos Pt$; $BP = O,\ B = O$

Substituting the values of integration constants into the general equation (3-7)

$$\varphi = \frac{b}{P^2}(1 - \cos Pt) \qquad (3.8)$$

Substituting the values of constants b and P into equation (5.8), we obtain the equation of motion of the moving system of the rock,.

$$\varphi = \frac{K[(l + r)\cos\beta - r] - 2Fl_1 - F_{\text{м}} \cdot l_{\text{м}}}{2Cl_1^2}(1 - \cos\sqrt{\frac{2Cl_1^2}{J}} \cdot t); \quad (3.9)$$

Solving the obtained equation of motion of the scalo with respect to K, we obtain the following calculation equation for determining the warp tension:

$$K = \frac{2Cl_1^2 \cdot \varphi + (2Fl_1 + F_{\text{м}}l_{\text{м}})(1 - \cos\sqrt{\frac{2Cl_1^2}{J}} \cdot t)}{[(l + r)\cos\beta - r] \cdot (1 - \cos\sqrt{\frac{2Cl_1^2}{J}} \cdot t)}; \qquad (3.10)$$

Since the tension studies in this chapter are carried out per machine cycle, the rolling angle of the scalo is constant, and the angle of convergence of the warp yarns from the winding of the warp β can be taken as zero, i.e. $\beta = 0$ equation (3.10) has the following form

$$K = \frac{2Cl_1^2 \cdot \varphi + (2Fl_1 + F_{\text{м}}l_{\text{м}})(1 - \cos\sqrt{\frac{2Cl_1^2}{J}} \cdot t)}{l \cdot (1 - \cos\sqrt{\frac{2Cl_1^2}{J}} \cdot t)}; \qquad (3.11)$$

The analysis of the formula shows that the filling tension (F) generated by the spring has a minimum value irrespective of the article of fabric to be produced (e.g. gauze, calico, tarpaulin, etc.). This tension is necessary to prevent shedding during shedding. The warp tension (F_M) created by the electromagnet will provide the technologically necessary force of the weft yarn. To illustrate the calculation of the warp tension for a weaving machine cycle with

by a micro gasket (STB). Let's assume the following initial data: *li* =15 cm;

l=15 см; l_M=15 см; r=6,6 см; F=17 кг; F_M=75 кг;J=15,3 кг.см.с^2; C=15,7 кг/см; φ= 0,035 рад; α=0°÷360°; n=180 мин$^{-1}$;

$$t = \frac{\alpha}{6 \cdot n}; \qquad P = \sqrt{\frac{2Cl_1^2}{J}} = \sqrt{\frac{2 \cdot 15.7 \cdot 15^2}{15{,}3}} = \sqrt{462} = 21{,}5$$

Table 3.2 shows the results of calculating the change in warp tension per revolution of the main shaft (cycle of operation) of the machine.

Table 3.2.

Results of calculating the change in warp tension per revolution of the main shaft (duty cycle) of the machine.

Angle of rotation of the main shaft, deg.	*cosPt*	*1-cos Pt*	*L(l-cos Pt)*	*(2Fli+F l_{MM}) (1-cos Pt)*	*K*, kg
0	*0*	1	15,0	510	50,5
10	*0,009*	0,991	14,87	505	50,6
20	*0,017*	0,983	14,75	501	50,7
30	0,026	0,974	14,61	1592	126,0
40	0,034	0,966	14,49	1579	126,0
50	0,043	0,957	14,36	1564	126,2
60	0,052	0,948	14,22	1540	126,4
70	0,06	0,94	14,1	1537	126,5
80	0,069	0,931	13,97	475	51,7
90	0,078	0,922	13,83	470	51,8
100	0,086	0,914	13,71	466	52,0
110	0,095	0,905	13,58	462	52,2
120	0,095	0,905	13,58	462	52,2
130	0,112	0,888	13,32	453	52,6
140	0,121	0,879	13,19	448	52,7
150	0,129	0,871	13,07	444	52,9
160	0,138	0,862	12,93	440	53,1
170	0,146	0,854	12,81	436	53,3
180	0,155	0,845	12,68	431	53,5
190	0,164	0,836	12,54	426	53,7
200	0,172	0,828	12,42	422	53,9
210	0,181	0,819	12,29	418	54,1
220	0,189	0,811	12,17	414	54,3
230	0,198	0,802	12,03	409	54,5
240	0,207	0,793	11,9	404	54,7
250	0,215	0,785	11,78	400	54,9
260	0,224	0,776	11,64	396	55,2
270	0,232	0,768	11,52	392	55,5
280	0,241	0,759	11,39	387	55,7
290	0,25	0,750	11,25	382	55,9
300	0,258	0,742	11,13	378	56,1
310	0,267	0,733	n,o	374	56,5
320	0,276	0,724	10,86	369	56,7
330	0,284	0,716	10,74	365	57,0
340	0,296	0,704	10,56	359	57,4
350	0,301	0,699	10,49	356	57,5
360	0,31	0,69	10,35	352	57,9

The table shows that the maximum warp tension corresponds to the moment of weft take-up (70° of machine main shaft rotation). After the weft break, the solenoid is deflected and the warp tension reaches its minimum value and fluctuates within small limits up to 20° of the machine main shaft position, i.e. up to the moment of the warp threads coming in. Then the electromagnet is switched on and the warp tension rises to its maximum value from 30° to 70° of the machine main shaft rotation. In order to ensure the normal weaving process, the warp yarns must have a certain filling tension. The filling tension (at closed shed) is necessary to create resistance for the warp yarns when the weft comes to the fabric edge and to ensure a clean shed opening. The filling tension varies depending on the fabric range, being higher for denser (heavy fabrics) and lower for less dense (light) fabrics. Incorrectly chosen filling tension causes violation of the technological process of fabric formation, changes its structure, increases the breakage of warp threads and reduces the efficiency of machine use and the quality of produced fabrics. Decrease of the filling tension causes decrease of warp tension at weft breakage, which increases the breakage stripe from cycle to cycle of the machine operation, weaving becomes impossible due to fabric stuffing. An increase in the filling tension leads to a decrease in the size of the surf strip, which can lead to warp overstressing. Consequently, a change in the value of the filling tension from the established norm causes disturbances in the technological process of fabric formation, changes its structure, increases the breakage of the warp yarns, and as a consequence reduces the quality of the fabrics produced and the productivity of the weaving machine. Therefore, the set filling tension of warp yarns for a certain article of fabric must remain constant during the whole period of winding operation on the weaving head. As follows from formula (3.11) the warp angle (β) of the warp yarns from the winding varies and can be determined practically. The same happens with the rolling angle of the scalo (φ). Therefore, it is to be expected that when the winding is actuated on the warp, there is an increase in the angles of convergence β and rolling . φ

Stabilisation of these parameters will lead to stabilisation of the warp tension when the warp winder is triggered. Here is a comparative calculation of the change in warp tension as the warp winding is triggered for the existing scalo system (without electromagnet) and for the modernised scalo system (with electromagnet). For the existing scalo system, the sum in the bracket ($2Fli+F\,l_{MM}$) remains constant, with F_M =0 during the machine cycle and over the entire winding actuation period. Consequently, the value of F increases by the value of F_M , i.e. the spring tightening increases. Let's assume the following initial data

For the existing scalo system:

l_1=15 см; l=15 см; r=6,6 см;

F=55 кг J=15,3 кг.см.с^2 C=15,7 кг/см;

φ= 120°; n=180 мин$^{-1}$; β=0-28°

t=0.11 сек; $1\text{-}cosPt$=0,905

For modernised scalo system (with electromagnet)

l_I=15 см; l=15 см; l_M=15 см

r=6,6 см; F=17 кг F_M=75 кг

J=15,3 кг.см.с^2 C=15,7 кг/см;

φ= 120°; n=180 мин$^{-1}$; β=0-28°

t=0.11 сек; $1\text{-}cosPt$=0,905

$$P = \sqrt{\frac{2Cl_1^2}{J}} = \sqrt{\frac{2\cdot 15{,}7\cdot 15^2}{15{,}3}} = \sqrt{462} = 21{,}5 \qquad 1\text{-}cosP\,t = 0{,}905$$

Table 3.3-3.6 shows the results of warp tension changes during winding operation on the weaving head for the existing scalo system (numerator values) and the modernised scalo system (denominator values). The analysis of Table 3.3 shows that the changes in the scalo angle and the yarn exit angle // as the winding is triggered on the winding head leads to an increase in warp tension. This is due to the change in the diameter of the winding on the warp, which leads: firstly, to an increase in the load on the warp threads due to an increase in the angle of thread run-off; secondly, to a change in the angle of deflection of the scalo, which leads to an increase in the spring tension. As a result, a constant warp release value is ensured by an increase in warp tension.

Table 3.3.

Results of warp tension variation at $\varphi \neq const$ $\beta \neq const$

Winding diameter per head, cm D

Диаметр намотки на навое, см D	55	45	35	25	15
φ, рад	0,035	0,044	0,052	0,061	0,07
$2Cl_1^2\varphi$	247	311	367	431	495
B, рад	0	7	12	17	28
$cos\beta$, рад	1	0,993	0,978	0,956	0,883
$(l+r)\cdot\cos\beta - r$	15,0	14,85	14,53	14,05	12,48
$(l+r)\cdot\cos\beta - r\cdot(1-\cos Pt)$	13,6	13,4	13,1	12,7	11,3
$2Fl_1\cdot(1-\cos Pt)$	1494/462	1494/462	1494/462	1494/462	1494/462
K, кг	128/52	135/58	142/63	152/70	176/85

Table 3.4.

Results of warp tension variation at $\varphi = const$ $\beta \neq const$

Winding diameter per head, cm D

Диаметр намотки на навое, см D	55	45	35	25	15
φ, рад	0,035	0,035	0,035	0,035	0,035
$2Cl_1^2\varphi$	247	247	247	247	247
B, рад	0	7	12	17	28
$cos\beta$, рад	1	0,993	0,978	0,956	0,883
$(l+r)\cdot\cos\beta-r$	15,0	14,85	14,53	14,05	12,48
$(l+r)\cdot\cos\beta-r\cdot(1-\cos Pt)$	13,6	13,4	13,1	12,7	11,3
$2Fl_1\cdot(1-\cos Pt)$	1494/462	1494/462	1494/462	1494/462	1494/462
K, кг	128/52	130/53	133/54	137/56	154/63

Table 3.5.

Results of warp tension variation at $\varphi \neq const \quad \beta = const$

Winding diameter per head, cm D

Диаметр намотки на навое, см D	55	45	35	25	15
φ, рад	0,035	0,044	0,052	0,061	0,07
$2Cl_1^2\varphi$	247	311	367	431	495
B, рад	0	0	0	0	0
$cos\beta$, рад	1	1	1	1	1
$(l+r)\cdot\cos\beta-r$	15,0	15,0	15,0	15,0	15,0
$(l+r)\cdot\cos\beta-r\cdot(1-\cos Pt)$	13,6	13,6	13,6	13,6	13,6
$2Fl_1\cdot(1-\cos Pt)$	1494/462	1494/462	1494/462	1494/462	1494/462
K, кг	128/52	133/57	137/61	142/67	146/70

Table 3.6.

Results of warp tension variation at φ=*const* β=*const*

Winding diameter at the head, cm D

Диаметр намотки на навое, см D	55	45	35	25	15
φ, рад	0,035	0,035	0,035	0,035	0,035
$2Cl_1^2\varphi$	247	247	247	247	247
B, рад	0	0	0	0	0
cosβ, рад	1	1	1	1	1
$(l+r)\cdot\cos\beta - r$	15,0	15,0	15,0	15,0	15,0
$(l+r)\cdot\cos\beta - r\cdot(1-\cos Pt)$	13,6	13,6	13,6	13,6	13,6
$2Fl_1\cdot(1-\cos Pt)$	1494/462	1494/462	1494/462	1494/462	1494/462
K, кг	128/52	128/52	128/52	128/52	128/52

At a constant scalo deflection angle (Table 3.5), the tension of the warp yarns increases as the winder is actuated due to the change in the angle of convergence β , of the winding yarns (Fig. 3.6). And for any diameter of winding D *of* the warp on the warp the value of deformation of the loading spring F *of* the mobile system of the scalo is constant. We have proposed a new system of warp tension release, which is presented in Fig. 3.6. At a constant angle of convergence β of warp yarns from the warp (Table 3.5), the tension of warp yarns increases by changing the angle of deflection φ of the scalo. And for any winding diameter on the warp, the angle of departure β of the warp yarns is constant. This variant is possible by installing an additional stationary roller in the warp-roll area (Fig. 6.2).

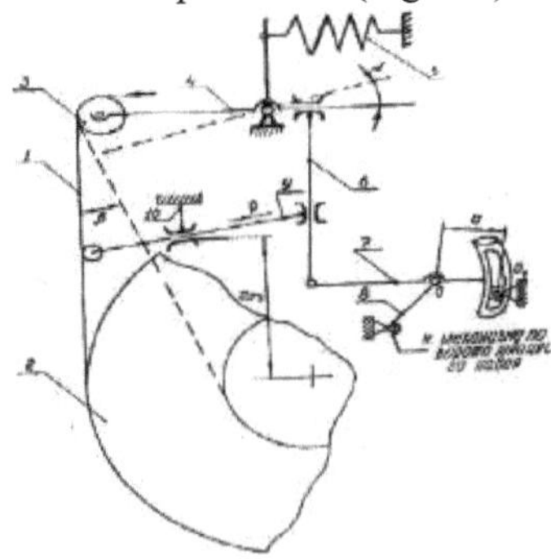

Fig. H.b. Variation of the warp tempering and tensioning system at the $\varphi = const, \beta \neq const.$

For the variant (Table 3.6) with a constant deflection angle φ and warp convergence angle β , the warp tension remains stable as the winding diameter D changes on the warp. This variant is possible with the complex use of devices according to a.c. No. 5014094 or patent No. 2016933 with a fixed additional roller in the nava- scal area (Fig. 3.7). In all variants, it follows from (Table 3.3-3.6) that the absolute values of warp tension are almost 2.5 times higher for the existing scalo system than the absolute values of the modernised scalo system. This positive property reduces the additional stresses on the warp yarns.

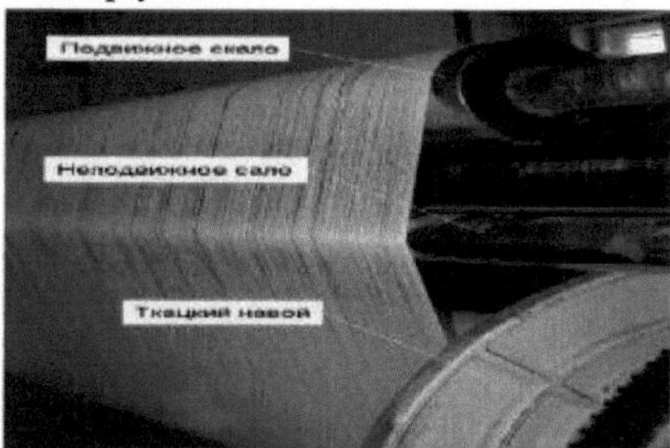

Fig.3.7. Variant of the warp tempering and tension system at *y^const const*

Figs. 3.8 - 3.11 clearly illustrate the changes in warp tension as the warp winding is triggered for the existing and upgraded scalo system.

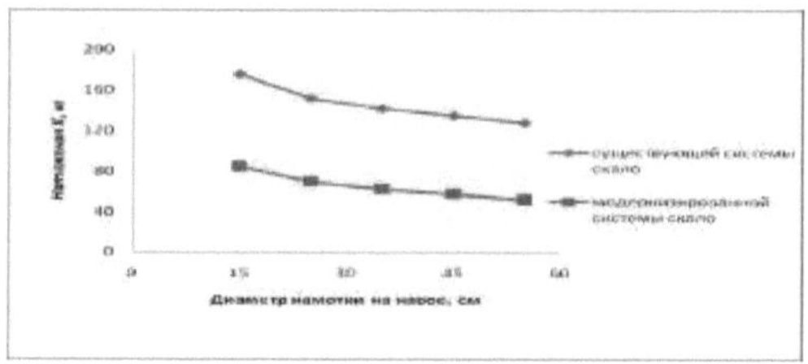

Fig.38.Plots of warp tension at $\varphi \neq const \quad \beta = const$

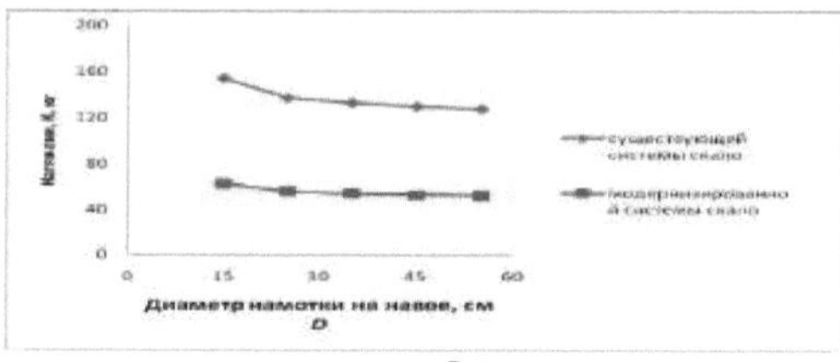

Fig. 3.9: Plots of warp tension at $\varphi = const, \ \beta \neq const.$

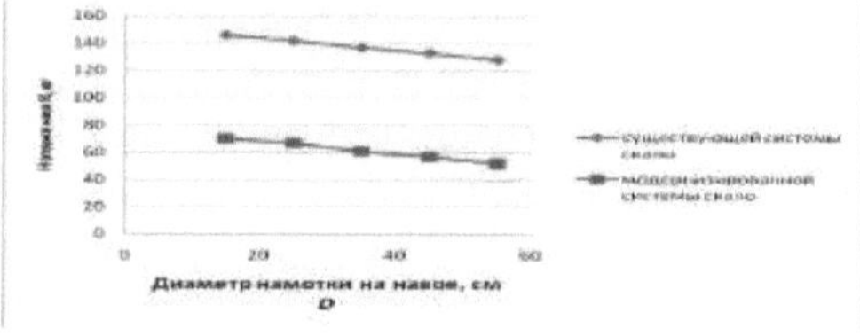

Fig.3.10. Plots of warp tension at $\varphi \neq const$, $\beta = const$.

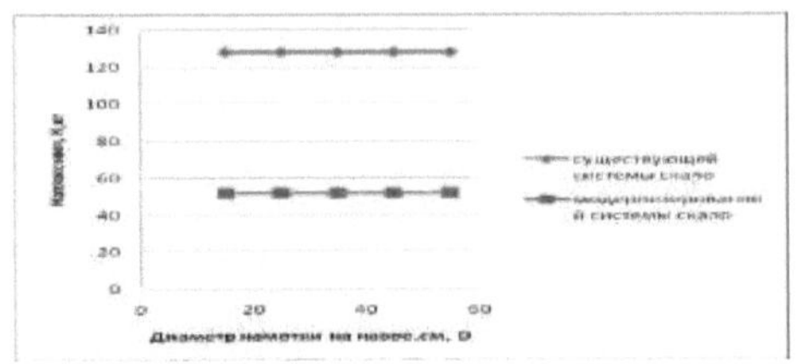

Fig.3.11. Plots of warp tension at $\varphi = const$, $\beta = const$.

The analysis of Table 3.3 shows that changes in the angle of rotation of the scalo a and the angle of convergence of the thread β during the winding operation on the warp leads to an increase in the warp tension by 36%. This is due to the change of the winding diameter on the warp, which leads: firstly, to an increase in the load on the warp threads due to an increase in the angle of the thread escape; secondly, to a change in the angle of deflection of the scalo, which leads to an increase in the stretch of the spring. As a result, the constancy of the warp release value is ensured by increasing the warp tension. At a constant angle of scalo deflection a (Table 3.4), the warp tension increases by 20% as the warp is triggered, due to the change in the angle of convergence /) of the winding threads. Moreover, for any diameter of warp winding D on the warp, the value of deformation of the loading spring F of the mobile system of the scalo is constant.

The changes in warp tension for the existing and new warp tempering and warp tension system are clearly illustrated in Fig. 3.12.

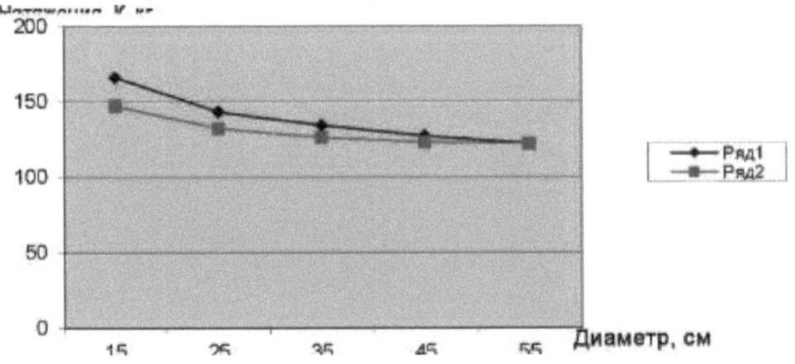

Fig. 3.12. Graphs of warp tension change as the winding is triggered on the warp for the existing and new tempering system. and warp tension.

The warp tension during loom start-up and shutdown has also been investigated. One of the main weaving defects that reduce the grade of wool fabrics is "start-up stripes" that occur during machine start-up. The reasons for the formation of start-up stripes can be the peculiarities of both the elastic system of the machine dressing and the design of the main components and mechanisms of the weaving machine involved in the fabric formation. On the weaving machine, the elastic filling system consists of two heterogeneous systems, warp and fabric, with different relaxation behaviour and during the machine stop, the relative warp and fabric lengths and the total tension of the elastic filling system change. At the same time, the total length of the elastic filling system remains unchanged. Consequently, the fabric edge (the warp-to-fabric transition boundary) moves either to the sternum or to the hems. To determine the displacement of the fabric edge, the experiments were carried out on the machines with microspacers (STB) filled with shirt fabric made of cotton yarn. The technological parameters of weaving are given in Table 3.7.

Table 3.7.

Technological parameters of weaving

№	Name	Unity of measurement.	Indicators
	Name of fabric		Shirt fabric
1	Machine type		STB-220
2	Margin value	hail	25
3	Yaw height	mm	100
4	Yaw depth	mm	145
5	Pharyngeal discharge	mm	440
6	Warp tension	gr.	25
7	Fibre type		Cotton
8	Linear density: weft warp	tex tex	15x2 15x2
9	Density: base to weft	thread/cm thread/cm	20 20
10	Intertwining		Canvas

The movement of the seam allowance during the machine standstill period was recorded using a developed device at a certain time interval. The machines were stopped in two positions: in the position of minimum warp tension (in the knot) and in the position of maximum shed opening (from the weft stylus). At the same time, the moisture content in the workshop varied from 54 % to 76 % for cotton yarn fabrics. Table 3.8 summarises the fabric downstitch movements for shirt fabrics. Table 3.8 shows that the down moves are more intensive during the first minutes of machine downtime. At that, the direction of the down displacement depends on the stop position of the main shaft of the machine. When the machine is stopped in the stagnation position, the fabric trim moves towards the sternum, and when the machine is started from this position, pronounced stripes (underslits) appear on the fabric. When stopping the machine at the position of the maximum open shed (260° - 280°), the fabric down moves towards the remise. In this case, the starting stripes are less pronounced or may be absent.

Table 3.8.

Moving the fabric edge for cotton yarn fabric.

№	*t*	Moving the fabric edge

	min	Machine standstill in a 20° jam			Machine stop at maximum opening shed of 250°		
		Moisture in the weaving department %					
		54	63	76	54	63	76
1	1	-0,011	-0,005	-0,04	0,012	0,004	0,050
2	2	-0,012	-0,006	-0,06	0,013	0,005	0,060
3	4	-0,013	-0,010	-0,08	0,014	0,010	0,090
4	6	-0,014	-0,013	-0,09	0,015	0,015	0,100
5	8	-0,015	-0,018	-0,100	0,016	0,020	0,110
6	10	-0,016	-0,025	-0,110	0,017	0,025	0,120
7	12	-0,017	-0,03	-0,120	0,018	0,031	0,126
8	14	-0,018	-0,035	-0,130	0,019	0,035	0,133
9	16	-0,019	-0,045	-0,135	0,020	0,045	0,139
10	18	-0,020	-0,055	-0,140	0,020	0,057	0,145
11	20	-0,021	-0,070	-0,150	0,020	0,068	0,150

The movement of the fabric edge towards the sternum is due to the fact that the warp yarns, being in the middle level, have a stiffness lower than the fabric stiffness, respectively, and the warp and fabric relaxation times. At machine stop 260° - 280° warp threads, deviating from the average level, change the geometry of threading, while the increase of warp stiffness is faster than fabric stiffness, respectively, warp and fabric relaxation times change. As the shop floor humidity increases, the magnitude of the downstitch movement increases during machine downtime. In addition, it is of interest to study the influence of the weaving machine type on the movement of the fabric warp in an elastic threading system. As each weaving machine is filled with the same fabric assortment, it has peculiarities of the filling - height and size of the shed, length of warp and fabric in the filling, etc. At the moment of machine stop, under the assumption that the movement of the fabric edge does not develop due to elastic deformation in time, we have:

$$X = \frac{K_O'}{C_O} - \frac{K_T'}{C_T}.$$

The deviation of the loom from the average level changes the filling tension (*KO*) of the warp yarns by AA$_0$, i.e.: ΔK_O, т.е: $K_O' = K_O + \Delta K_O$. On the other hand, the changes in warp tension can be expressed by the product of warp deformation by the warp stiffness coefficient (C_o), taking into account the coefficient (*n*) of the ratio of shed height to warp length in an elastic weaving machine dressing system: $K_O' = K_O + \Delta\lambda_O \cdot C_O \cdot n$

The deformation of the warp yarns when they deviate from the centre line has the form

$$\Delta\lambda_O = \frac{h^2}{2} \cdot \left(\frac{1}{l_1} + \frac{1}{l_2}\right)$$

where: *h* - deviation of the warp threads from the centre line of the machine threading (half of the shed height from AND); $/_7$ - length of the front part of the shed; l_2 - length of the back part of the shed.

After substitution we obtain

$$X = \frac{K_O + \frac{h^2}{2}(\frac{1}{l_1} + \frac{1}{l_2}) \cdot C_O \cdot n}{C_O} - \frac{K_T}{C_T}. \quad (3.12)$$

The analysis of formula 3.12 shows that when the weaving machine stops, the movement of the fabric edge (*X*) can have the following sign values: at zero value (A), the fabric edge is stationary, we can expect a defect-free weaving; at minus value (A), the fabric edge moves towards the sternum, we can expect a defect in the fabric "starting understitch", i.e. underdensity in the weft; at plus value (A), the fabric edge moves towards the remise, we can expect a defect in the fabric "starting blockage", i.e. an underdensity in the weft.i.e. underestimated fabric density in the weft; at the plus value (A), the fabric down moves towards the remise, we can expect the defect on the fabric "start-up underseeka", i.e. overestimated fabric density in the weft. The calculations of movement of the fabric down during the machine stop at different types of equipment tucked fabric "Sorochechnaya", which is produced by plain weave with tucking tension of warp and fabric $K_o = K_T = 20$ *sNlinear* density of the main thread $T_O = 15x2$ *tex*, linear density of weft yarn $T_y = 15x2$ *tex*, *with* stiffness coefficient of a metre section of warp yarns $C'_o = 1,0 \frac{H}{мм}$ and stiffness coefficient of fabric in terms of a single thread of a metre section. $C'_T = 0,3 \frac{H}{мм}$.

Table 3.9.

Filling parameters for weaving machines

№	Name of indicators	Weaving machine type					
		Pnev-mora-pirny, ATPR	With micro-pro-puters, STB	Rapier, P-190 Zul-cer- ryuti.	Rapier, Somet Super Excel	Shuttle, AT	Pneumatic, Toyota JAT 810
1	Socket height *H, mm*	50	60	60	80	100	80
2	Front part of the yawn *h, mm*	80	150	170	150	230	150
3	Posterior part of the yawn *h, mm*	400	350	500	500	420	500
4	Total length of warp yarns in the filling $L_{,O}$ *mm*	1700	1900	1900	2000	1600	2000
5	Total length of fabric in the filling L_T, *mm*	600	600	600	500	500	500
6	Stiffness coefficient of a single warp yarn C_o, $\frac{cH}{мм}$ at the petrol station	59	53	53	50	63	50
7	Fabric stiffness coefficient in	50	50	50	60	60	60

	the dressing in terms of single ... g^{cH} warp yarn $C_T, \frac{cH}{мм}$						

Table 3.9 shows the filling parameters and the results of the fabric warp movement calculation, Table 3.10 the results of the fabric warp movement calculation for different types of looms, Table 3.11 the effect of the filling tension of different types of weaving machines on the fabric warp movement. Analysing the tables shows that the direction of the fabric edge movement (*X*) depends on the stop position of the machine main shaft.

Table 3.10.

Calculation results of the fabric edge movement for different machine types

№	Deviation of warp yarns from the centre line *h, mm*	Traversing of the fabric edge at machine standstill *X,mm*					
		Weaving machine type					
		Pnev-mora-pirny, ATPR	Microlayers, STB	Rapier P-190 Sulzer-Ruti.	Rapier, Somet Super Excel	Shuttle, AT	Pneumatic, Toyota JAT 810
1	0	-0,061	-0,024	-0,023	0,066	-0,016	0,066
2	5	-0,06	-0,022	-0,022	0,068	-0,015	0,068
3	10	-0,057	-0,020	-0,021	0,070	-0,013	0,070
4	15	-0,046	-0,014	-0,016	0,074	-0,09	0,074
5	20	-0,026	-0,003	-0,006	0,084	0,001	0,084
6	25	0,007	0,016	0,010	0,101	0,017	0,101
7	30	-	0,045	0,033	0,125	0,040	0,125
8	35	-	-	-	0,160	0,073	0,160
9	40	-	-	-	0,206	0,117	0,206
10	45	-	-	-	-	0,173	-
11	50	-	-	-	-	0,244	-

Table 3.11.

Influence of warp filling tension, different types of weaving machines of different types of looms on the movement of the fabric warp

№	Filling tension of warp yarns, cN.	Traversing of the fabric edge at machine standstill *X,mm*					
		Weaving machine type					
		Pnev-mora-pirny, ATPR	Micro gaskets, STB	Rapier, P-190 Zul-cer- ryuti.	Rapier, Somet8 uper Excel	Shuttle, AT	Pneumatic, Toyota JAT810
1	5	-0,015	-0,006	-0,006	0,017	-0,004	0,017
2	10	-0,031	-0,011	-0,011	0,33	-0,008	0,33
3	15	-0,046	-0,017	-0,017	0,050	-0,012	0,050
4	20	-0,061	-0,024	-0,023	0,066	-0,016	0,066
5	25	-0,076	-0,028	-0,028	0,083	-0,020	0,083

6	30	-0,092	-0,034	-0,034	0,100	-0,024	0,100
7	35	-0,107	-0,040	-0,040	0,117	-0,028	0,117
8	40	-0,122	-0,045	-0,045	0,133	-0,032	0,133
9	45	-0,137	-0,051	-0,051	0,150	-0,040	0,150
10	50	-0,153	-0,057	-0,057	0,167	-0,40	0,167

On air-jet sawing machines (ATPR), with microlayers (STB), rapier weaving machines (R-190) and shuttle weaving machines (AT), when the machine is stopped in the backstop position ($h=0$), the fabric edge moves towards the sternum and when the machine is started from this position, stripes appear on the fabric (starting underslit). This defect is particularly pronounced on the ATPR weaving machine. On rapier (Somet SuperExcel) and pneumatic (Toyota JAT 810) machines, the movement of the fabric edge moves towards the remise, which causes stripes (start-up sap). Increase of the shed height causes the transition of the fabric from the starting undercut to the starting slaughter or strengthens the starting slaughter (Somet SuperExcel and Toyota JAT 810). By adjusting the warp and fabric stiffness coefficients, it is possible to slightly reduce the starting slack, as the total length of the warp yarns in the filling is unstable and depends on the diameter of the warp winding on the warp. In addition, the absolute values of the defect values "starting stripes" increase with increasing filling tension of the warp yarns. Therefore, it is reasonable to produce fabrics with minimum filling tension and maximum warp tension at the moment of weft surfing. This is possible by using special means that relieve the elastic filling system when the machine is stopped and load the elastic filling system when the weft yarn breaks. In most cases, the machine stops in the position of the backstitch, therefore, according to the results obtained, the movement of the down is towards the sternum, and the fabric is understitched. In another variant the system of automatic unloading of the elastic filling system tension during the machine stoppage (idle time) and its loading during the machine start-up period is offered, in which after oscillograms processing the following results of the warp filling tension values are obtained are presented in table 3.12. As it is seen the new system unloads the elastic filling system during the machine stoppage (idle time) by 35 % and restores it to the required value before the weft filling during the machine start-up period.

Table 3.12.

Results of warp filling tension values

№	Name	Filling tension of warp yarns , cN	
		The existing rock climbing system	A new rock climbing system
1	Machine shutdown (downtime) period	34	24
2	Machine start-up period	34	33
3	Machine operating period	33	20

Toyota machines are equipped with a system for preventing the formation of starting stripes, which guarantees high fabric quality. The system of preventing the formation of starting fringes (Fig. 3.13) has: electronic warp release 1 and electronic fabric diversion 2, which relieve the elastic filling system when the machine stops and load the elastic filling system when the filling yarn is picked up 3; the starting mode of the main motor guarantees a full pick-up force already at the first weft pick-up.

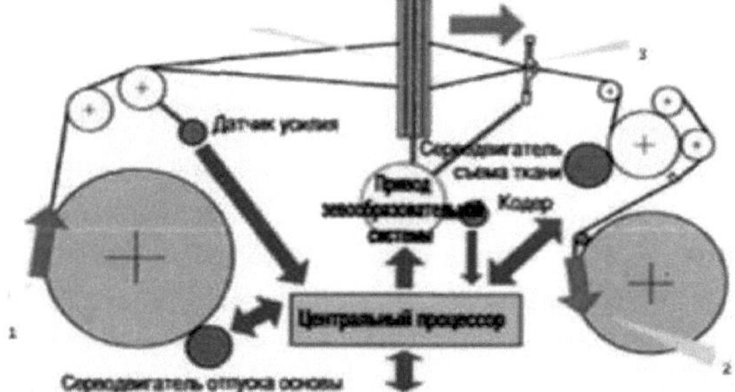

Fig. 3.13. System of preventing the formation of start stripes on the fabric, where 1- electronic warp release, 2- electronic fabric diversion, 3- weft yarn breakage.

Forward fabric down - lowering the warp tension immediately after machine stop prevents the fabric down and the reed from coming into contact, thus eliminating the cause of start-up stripes (Fig. Fig. 3.14). After starting the machine, the programmed tension is automatically restored and the surfing takes place at the normal position of the fabric warp. Warp release setting - the operator can set the permissible amount of warp tension reduction in case of machine stoppage or idle time to reliably eliminate the formation of start-up bands. Single weft laying - single weft laying can be performed without reed surfing during machine start-up. This function is particularly useful for avoiding the formation of start-up stripes during the production of fabrics.

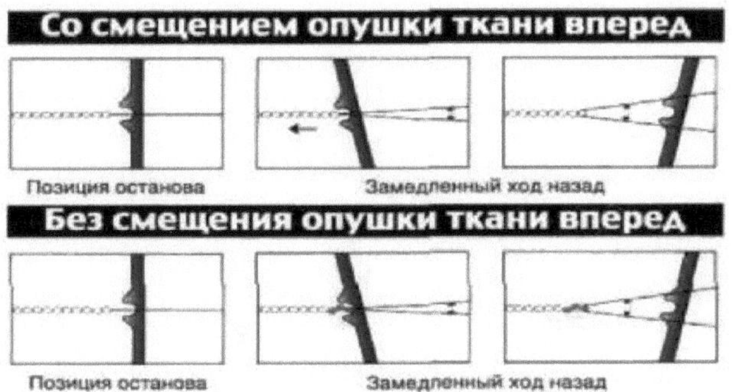

Fig. 3.14 Interaction of the reed with the fabric warp during weft inflow. Selecting the stop and start angle of the machine - by programming the desired stop and start angle according to the fabric type, the weaver can prevent the formation of starting stripes.

The fourth chapter contains technological and consumer research of shirt fabrics and optimisation of its production technology. Reducing the level of thread breakage in the weaving process to the greatest extent (at a constant speed of the main shaft of the machine) contributes to the achievement of the goal. This allows us to choose as the main criterion for optimisation of the weaving process the minimum breakage of main and weft yarns. On the one hand, they determine such indicators as weaver's workload, equipment utilisation rate, which significantly influence the labour and equipment productivity. On the other hand, the breakage rate can unambiguously determine the conditions of fabric formation on the weaving machine depending on such technological parameters as filling tension of warp and weft yarns, which can be determined by the minimum value of thread breakage. Two main parameters, i.e. two main independent variables were selected: x 1 - filling tension of weft, cN; x_2 - filling tension of warp, cN. It is established that warp thread breakage at low values of filling tension increases at the expense of increased surf. Then, as the warp tension increases, the breakage rate decreases and increases again as the warp tension increases further due to overstressing of the warp yarns. The selected factors meet all the requirements of the theory of mathematical planning of experiment: there is no interchangeability of factors, they can be measured by available means, they can be varied within a sufficiently wide range of minimum and maximum values and take them with the necessary accuracy. As for the rest of the technological parameters of machine fuelling, all of them were constant during the experiment. In this work, a second-order central composite method of experiment planning was adopted, which provides an opportunity for detailed study, description and optimisation of the weaving process in the optimisation domain under investigation. The selection of intervals and values of factors for five levels of variation was carried out taking into account the technological possibilities of loom fuelling. (Table 4.1)

Table 4.1.

Factors	Levels of variation	Interval

	-1,414	-1,0	0	+1,0	+1,414	
xi- refuelling tension of weft yarns, cN.	3	5	10	15	17	5
X2-fuelling tension yarn bases, sN.	13	15	20	25	27	5

Table 4.2.

RCCE planning matrix

No. of Experience	X1	X2	*x1*	X1X2	*X1*	YU	YR	$(Y_R-Y_U)^2$
1	-	-	+	+	+	0,32	0,33	0,0001
2	+	-	+	-	+	0,34	0,34	0
3	-	+	+	-	+	0,35	0,34	0,0001
4	+	+	+	+	+	0,36	0,36	0
5	- 1,414	0	2	0	0	0,32	0,31	0,0001
6	+1,414	0	2	0	0	0,34	0,33	0,0001
7	0	- 1,414	0	0	2	0,36	0,35	0,0001
8	0	+1,414	0	0	2	0,39	0,37	0,0004
9	0	0	0	0	0	0,29	0,30	0,0001
10	0	0	0	0	0	oh,h	0,30	0
11	0	0	0	0	0	0,31	0,30	0,0001
12	0	0	0	0	0	0,30	0,30	0
13	0	0	0	0	0	0,300,	0,30	0

The experiment conducted on the selected matrix allows to obtain a second-order mathematical model describing the influence of factors xi, X2 on the selected optimisation parameters of the following form

$$y = в_0 + в_1 x_1 + в_2 x_2 + в_{12} x_1 x_2 + в_{11} x_1^2 + в_{22} x_2^2$$

where in_0 , e_i , *vu, v"* - regression coefficient; in_0 - free term; e_i -1, 2, - regression coefficients at linear terms; *vu=1*, 2, - coefficients at interaction of factors; *v"* = - *regression* coefficients of squared terms.

An adequate mathematical model describing the dependence of cliffiness on the selected significant factors has been obtained, it has the following form

$$Y_R = 0,3 + 0,008X_1 + 0,011X_2 - 0,003X_1 \cdot X_2 + 0,011X_1^2 + 0,34_2^2 \qquad (3.1)$$

The most efficient evaluation of the technological experiment can be carried out by means of cut-offs : y= f (xi) at constant X2 ; y= f (xz) at constant x_1 . Tables 4.1-4.2 summarise the results of the cut-off calculations from the input factors. As can be seen all equations are parabola equation. Analysing the curves in Figs. 4.1-4.2 plotted by the obtained equations shows that the change of y from x_1 and X2 has the form of concave parabolas. At zero value of the filling threads' filling tension x_1 and at zero value of the main threads' filling tension X2 it is possible to reduce the breakage of weft and main threads by 37%, i.e. the parameters will have the following values: weft threads' tension -10 cN; main threads' tension - 20 cN (per 1 thread). At these values of parameters the breakage of weft threads and main threads will not exceed 0.3 breaks per 1 metre of fabric.

Table 4.3.

Calculation results $y = f(x_1)$ at constant X2

№	Constant values of factors	Breakage of warp yarns at factor value xi				
		-1,414	-1	0	1	1,414
1	X_2 = -1.414	0,359	0,353	0,354	0,378	0,394
2	X_2 = -1	0,331	0,325	0,325	0,347	0,363
3	X_2 = 0	0,313	0,305	0,300	0,321	0,335
4	$X = 1_2$	0,362	0,353	0,347	0,363	0,376
5	X_2 = 1.414	0,402	0,393	0,386	0,4	0,413

Table 4.4.

Results of calculation y = f (X2) at constant x_1

№	Constant values of factors	Breakage of warp yarns at factor value xg				
		-1,414	-1	0	1	1,414
1	X1 = -1,414	0,359	0,331	0,313	0,362	0,402
2	X1 = -1	0,353	0,325	0,305	0,353	0,393
3	X1 = 0	0,354	0,325	0,300	0,347	0,386
4	X1 =1	0.378	0,347	0,321	0,363	0,4
5	X1 =1,414	0,394	0,363	0,335	0,376	0,413

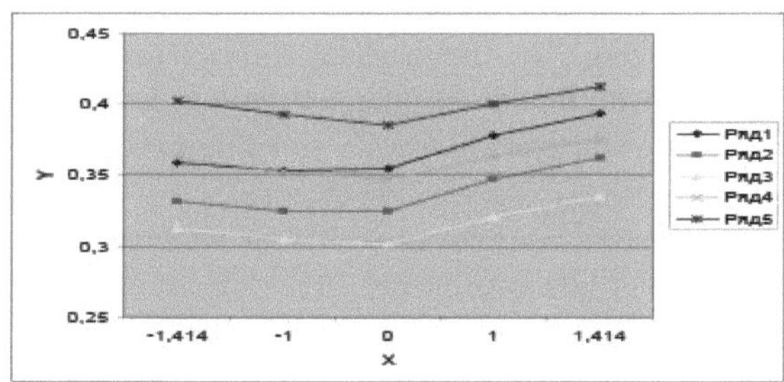

Row1 X2=- 1.414; Row2 X2= -1; RowZ xg= 0; Row4 xg = +1; Row5 X2= +1.414; Figure 4.1.

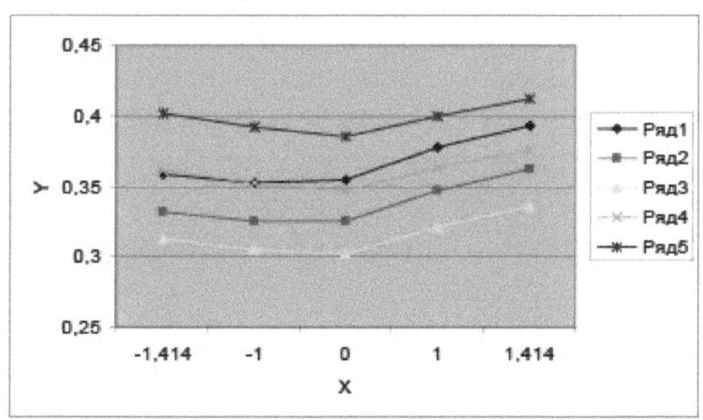

Ряд 1 x_1 = -1,414; Ряд 2 x_1 = -1; Ряд 3 x_1 = 0;
Ряд 4 x_1 = +1; Ряд 5 x_1 = +1,414;
Рис. 4.2.

Row1 xi = -1.414; Row2 xi = -1; RowZ xi = 0;
Row4 xi = +1; Row5 xi = +1,414;
Figure 4.2.

In order to obtain a shirt fabric of a certain structure with certain properties, it is necessary to create mutual pressure forces between the threads of the fabric. Mutual pressure forces between the fabric threads are created in the process of its formation on the weaving machine and determine the mutual arrangement of the fabric threads, the latter depends on the following factors: the type of raw material used; the diameter of the main and weft threads and their ratios; the density of the fabric in warp and weft and their ratios; the type of weave of the fabric; the tension of the main and weft threads and the ratio of tension; technological parameters of dressing and production of the fabric, affecting the tension of the main and weft threads. The type of raw materials for shirt fabric is selected taking into account the purpose of the fabric and the requirements to it. The properties of the yarns used in warp and weft largely determine the properties of the fabric made of them. Changing the type of raw material in at least one system of yarns in the warp or weft of the fabric has a significant impact on the technological parameters of its production, fabric structure and properties. Thus, with a change in the type of raw material in the weft, other things being equal, the tension of the main threads changes at the moment of surfing, i.e. at the moment of formation of a fabric element. Fig. 4.3 shows the oscillogram of cotton main threads tension when producing fabric on the machine with microspacers (STB) by plain weave Ro=Ru=2O n/cm with the use of different types of raw materials in the weft. The oscillograms of the main threads tension at fabric production with the use of different types of raw materials in the weft are obtained.

The oscillograms show that the warp yarns of different weft types experience different tensions at the moment of the interference. The highest tension at this point is experienced by the warp yarns when capron, polypropylene and nitre yarns are used in the weft.

The warp yarn tension is higher at the yawing moment than at the run-in moment, but its value is approximately the same for all raw materials in the weft. At the moment of formation of the fabric element, i.e. at the surf, the warp yarns experience the highest tension, as the warp tension is at its maximum at this moment. Moreover, when capron polypropylene and nitron weft yarns are used, the tension of the warp yarns is the highest at the moment of surfing. This indicates that the distinctive features of the properties of these weft yarns influence the technological parameters of fabric production on the weaving machine. The change in the type of raw material of the weft yarns influences the structure and properties of the fabrics produced. The change in the fabric structure can be characterised by the weft and weft yarn yields and the warp and weft densities.

Table 4.5.

Results of research on the properties of weft yarns.

№	Name of properties of weft yarns	Properties of weft yarns		
		Initial module stiffness, kgf/mm	Bursting load, kgf	Rupture elongation %

1	Lavsan T=15x2 tex	6,4	1,4	13,7
2	Nitron T=15x2tex	5,6	1,0	7,6
3	Capron T=15x2 tex	4,5	1,8	13,6
4	Polypropylene T=15x2 tex	2,9	1,5	20,3
5	Cotton yarn T=15x2 tex	1,5	oh,h	6,8

Table 4.6.

: Tissue findings.

№	Name of weft properties threads	Indicators of fabric properties (linear density on the base To=15x2 tex. basis density Ro=200 yarns/dm.)					
		Bursting load, kgf		Breaking elongation %		Yarn processing in fabric %	
		basis	ducks	basis	ducks	basis	ducks
1	Lavsan 15x2 tex	28	120	23	29	7	2
2	Nitron 15x2tex	29	114	29	31	10	3
3	Capron 15x2 tex	26	162	15	51	5	4
4	Propylene 15x2 tex	25	140	26	52	9	5
5	Cotton yarn 15x2 tex	31	24	9	10	7	7

It follows from Tables 4.5 and 4.6 that with increasing modulus of stiffness of weft yarns, weft yarns yield decreases. Cotton weft yarns in the fabric have the highest yield value. This is explained by the fact that the smallest value of the initial modulus of stiffness of cotton yarns at the moment of formation of a fabric element allows to

The weft yarns are easily deformed. Analysing the breaking load data in Table 4.6, we can conclude that fabrics with cotton yarns in the weft have a higher warp direction breaking load. This indicates that the breaking load in the warp direction is influenced by the breaking load of the yarns used in the warp and the coefficient of friction between the yarns in the fabric. All the tested samples have cotton yarns in the warp, and since the coefficient of friction between the yarns in cotton fabric is the highest, the fabric using cotton weft has a higher breaking load in the warp direction. The breaking elongation in the warp direction is different for all fabric samples: the lowest is in samples with cotton yarns in the weft; the highest is in samples with nitron yarns in the weft. This indicates that changing the type of raw materials in the weft causes a change in the fabric properties not only in the weft direction, but also in the warp direction. The conducted researches allowed to establish that technological parameters of fabric production, structure and properties depend on the properties of yarns used in the weft. Therefore, when designing a fabric with predetermined properties it is necessary to pay special attention to the properties of warp and weft yarns. The samples of the produced fabrics were tested for physical and mechanical properties in the laboratory of the certification centre "CENTEX UZ on the following indicators: tests for fabric resistance to abrasion; tests for air permeability; tests for breaking load and elongation. The results of the tests are given in Table 4.7.

Table 4.7.

№	Name	Unit.	Basis	Duck	Fabric

1	Linear density	tex	15x2	15x2	-
2	Fabric density	yarn/dm.	200	200	-
3	Surface density fabrics	gr/m^2	-	-	119
4	Air permeability	cm^3/cm^2 sec.	-	-	86
5	erasure	cycle	-	-	18000
6	Breaking load	H	307	220	-
7	Breaking elongation	%	7,8	7,8	-
8	Yarn processing in woven fabrics	%	6,6	7,0	-

CONCLUSION

Taking into account climatic conditions of the region it is necessary to produce shirt fabrics with surface density up to 100 g/m^2 . And the technology of shirt fabrics manufacturing is carried out without the process of sanding, by using twisted yarns of low linear density. On the basis of the theory of wave propagation in elastic medium it is established that the cause of weft thread breaks is the impact of the brake foot on the weft thread, at increasing the speed of laying, the deformation and tension of the weft thread increase on average by 30% and at increasing the coefficient of friction of the thread on the surface of the laying mechanisms, the tension and deformation of the weft thread decrease by 19%. A new system of weft braking and feeding is developed and tested. The regularities of change of the sinking string tension depending on the friction radius, friction angle, friction coefficient, rigidity of the sinking string and compensator position have been obtained. A stand and methodology for determining the friction coefficient in the compensator eye, depending on the friction radius, type and linear density of the thread, type of the compensator working surface are developed. Theoretical expert studies of warp thread tension for the existing and new system have been carried out, where the expediency of using the new system of warp thread tension regulation during the machine cycle, as the warp is actuated and during the machine start-stop period has been shown. The technological process of production on a weaving machine has been investigated by means of the mathematical method of second-order rototable planning of experiment. The geometrical interpretation of the mathematical model is studied by means of slices. The optimum technological parameters of shirt fabric production are determined, where the thread breakage is 0.3 breaks per 1 m of fabric at weft thread tension -10 cN and warp thread tension-20 cN. It is determined that technological parameters of production, structure and properties of shirt fabric depend on the properties used in weft yarns. With the increase of the stiffness modulus of weft yarns, the weft yarns processing decreases. With the use of cotton weft, the fabric has a high breaking load in the warp direction, due to the high coefficient of friction between the warp and weft yarns. The breaking elongation in the warp direction for all fabric samples is different: the lowest in the fabric with cotton yarns in the weft is the highest in the fabric with nitron yarns in the weft.

LITERATURE

1 .HANDB0OK OF WEAVING Edited by S Adanur, Department of Textile Engineering, Auburn University, USA 440 pages 543 figures 68 tables 254 x 176mmhardback2000 .

2 .HANDBOOK OF YARN PRODUCTION Technology, science and economics P R Lord, NCSU, USA 504 pages 244 x 172mm hardback July 2003.

3 .Talavashek O. et al. "Needleless weaving machines". Moscow, Legprombytizdat 1985.

4 . Borodin. A.I. et al. Reference book "Filling calculations of sulphur fabrics" I II ch. 1970 г.

5 . Bukaev P.T. Reference book on cotton weaving. M.1979.

6 Damyanov G.B., Bachev C.Z., Surnina N.F. Fabric structure and modern methods of its design. -M.: Light and Food Industry, 1984.

7 .Rakhimkhodjaev S.S., Kadyrova D.N. "Modern methods of fabric design", Tashkent. TITLP.2006

8 Sklyannikov V.P., Mashkova E.N. Research of influence of fabrics from chemical fibres on their air permeability // Textile industry. 1973.№6.C. 75.

9 Vishnevskaya L.I. Research of influence of fibre property and structure on operational properties of multi-component fabrics: Abstract of thesis. M.,1977

10 . Margolin I.S. Wear resistance of fabrics from wool and chemical fibres. M., 1967.

I.Eremina N.S. Study of the regularity of changes in physical, mechanical and hygienic properties of fabric from its structure. M., 1952.

12 Martynova A.A., Slostina G.L., Vlasova P.A. Structure and design of fabrics. M., RIO MGTA, 1999.-434c

13 . Martynova A.A. Factors influencing the structure and properties of tissues. M., 1976.

14 Arkhangelsky N.A. Air permeability of fabrics depending on their structure // Textile industry. 1949. №7.C.25-27.

15 Vorobyev V.A. Calculation method for construction of woollen yarn and fabric. M. 1964.

16 Denisenko T.N. Development of methods for estimation of tension of weaving machine dressings. Dissertation abstract. -M.,1993.

17 Bubentsov L.V. Influence of density on warp and weft and weave on static electricity charge potential // Izv. of higher educational institutions. Technology of textile industry. 1978. №11.C.38-40.

18 Sklyannikov V.P. Methods of experimental determination of the order of the structure phase of plain weave fabrics // Izv. of universities. Technology of textile industry, -1967.- No.1,- P.20-24.

19 .Novikov N.G. About structure of fabric designing it with the help of geometrical method Textile industry. 1946 №2,4,5,6,11.C.42

20 .Urazov N.H. To the methodology of fabric design Textile industry. 1968.№7.

21 Kutepov O.S. Methodology of designing fabrics by the given weight of a square metre // Textile Industry,-1950,-¹2
22 Kuznetsov A.M. About designing of fabrics. Textile industry. - 1951.-№7
23 Rachenkova O.M. Development of the methodology of calculation of rational parameters of structure of fabrics of different weave taking into account the technology of their manufacturing. Dissertation abstract. - M., 2000.
24 Damyanov G.B., Bachev C.Z. Fabric structure and modern methods of its design. M., 1984.
25 N.F. Surnina N.F. Designing of fabric according to the given parameters. M., 1973.
26 Gordeev V. A. Research of the mechanisms of tempering and warp tensioning of weaving machines: Dissertation... Doctor of Technical Sciences, - M., 1953, -320 p.
27 Gordeev V.A. Dynamics of mechanisms of tempering and warp tensioning of weaving machines, - M.: Light Industry, 1965.-227 pp.
28 Voronina E.A. About regulation of warp tension in the weaving machine cycle. N.tr., VNIILtekmash, 1957, No. 2. Research of weaving machines, p.171 -180.
29 . Pfohl Walter. Der bewegliche streichbaum als Ausgleichsfaktor von Spannungs differenzen. Milland Textil berichte, 1953, no. 9.
30 Kolesnikov P.A. The tension of main threads in the process of weaving and its influence on physical and mechanical properties and breakage of warp threads: Dissertation...candidate of technical sciences, - M., 1949, -418s.
31 .Brokel Gerchard. Die Verandenung der Kettpadenspannung bei Baum Wollwebstuhlen min dem Kettverlauf und der Schaftsahl.TextilPraxic, 1961,No. 6.
32.Erokhin Y.F. Research and improvement of weaving process in cotton production: Dissertation...doctor of technical sciences, - M., 1980, - 242 p.
ZZ.Ohunboboev O.A., Ergashov M. Theory of calculation of warp tension in silk weaving machines. Fan va technology. Tashkent. 2010. 224 c.
34.Ohunboboev O.A.. State of the question and improvement of mechanisms of a shuttleless weaving machine for production of natural silk fabric. Fan va technology. Tashkent. 2016. 128 c.
35 .Bukaev P. T. Optimisation of the weaving process on needleless looms. M. Legprombytizdat, 1990. 176 c.
36 Drohlyansky I.M. Theoretical and experimental study of the elastic system of the STB machine dressing during the production of multilayer woollen fabrics: Dissertation.... Candidate of Technical Sciences, - M., 1970, - 252 pp.
37 Vlasov P.V. Normalisation of weaving process. - M.: Light and Food Industry, 1982.-296 pp.
38 Khamraeva S.A. Increase of wear resistance of fabrics by optimisation of parameters of their formation. Diss....doct. of technical sciences,- Tashkent, TITLP, 2010.
39 Erokhin Y.F. Movable system of the air-jet weaving machine scala,- Izv. of high schools. Technology of textile industry, 1971, № 4, p.93-95.

40 Optimal parameters of installation of mechanisms of tempering and tensioning of weft and warp on machines STB-2-330 SHL / A.I.Makarov and others,-. M.:TsNIITEIlegprom, 1973,- 48 p.
41 .Greenwood by K, Cowhing W.T.. The position of the cloth felle in power looms part I-stable weaving condititions, part Il-Disturbed weaving conditions, part Ill-Experimental. Journal of the Textile Institute, Fransaetions,1956,No. 5.
42 Balod J. Ya., Milasius V.M. Investigation of relaxation and anstringing phenomena on a pneumatic machine,- In book: Materials of the Lithuanian Republican XXII Scientific and Technical Conference. Kaunas, 1972.
43 Petukh N.A. Investigation of the reasons for the appearance of start stripes in the fabric produced on ATPR machines. Nauch.tr. TSNIIHBI, 1977, NO.1. Questions of new technology in the cotton-paper industry, pp. 119-122.
44 Bukaev P.T. Kinematics of weft yarn surfing and its influence on weaving process on shuttleless weaving machines, - Scientific tr. TSNIIHBI, 1977, NO. 1. Issues of new technology in the cotton industry, p.113-118.
45 .Zhao Jian. To the question about the movement of the fabric edge at stopping of the weaving machine. - Izv.vuzov. Technology of textile industry, 1961, No.1, p.71-79.
46 Burnashev R.Z. Research of the surfing process on weaving machines: Cand. Candidate of Technical Sciences. - M., 1969. - 196 c.
47 .0rnatskaya V.A. et al. Designing and modernisation of weaving machines. M., L.I., 1986.
48 .Kuzovkin K.S. et al. Experience of work on STB machines., M. L.I. 1968.
49 D.N. Research of weft yarn tension on machines with microspacers. T.P. 1965, No. 8.
50.Topilin A.P. et al. Automatic shuttleless weaving machines STB 2-250.,M.,L.P., 1969.
51.Shamshtein A.I. Research of the process of winding textile threads from the surface of conical chucks made of different materials. Dissertation abstract., Kostroma, 1973.
52.Efremov E.D. About non-uniformity of movement of a thread at rewinding on winding machines. Autoreferatdissertation., Ivanovo, 1963.
53.Efremov E.D. On the influence of guiding devices on the tension of the moving thread., T.T.P., I960, No. 1.
54.Blinov I.P. Improved weft brake of the STB machine. T.P., 1982, No.11.
55.Stolyarov A.N., Abramov V.A. Modernised weft brake of the STB machine. R.S. Weaving, 1983, No. 44.
56.Petukhov V.A. Modernisation of shuttleless weaving machines. Kiev, Technics, 1987.
57.Biconcini G. Rend. R. Accod. d. lincei, 6a, 1925.
58 .Lorenz H. Technishe Meeh. Starrer Gebilde, 1924.
59 Minakov A.I. Fundamentals of yarn mechanics. MTI. M. 1941.
60 .Urazbayev M.T. Fundamentals of mechanics of weighted deformable flexible

thread.Tashkent,1951.
bEalimova H.A. Waste-free technology of silk processing. Tashkent. Fan, 1994, 310 pp.
62 . Alimova H.A. et al. Rakhmatullin waves in filaments and rods. Tashkent, Fan, 2001.
63 Ergashev M. Properties and interaction of waves in a filament. Tashkent, Fan, 2001.
64 Ergashev M. Wave problems of impact of a thread with solid bodies. Tashkent, Fan, 2001
65 Rasulov X. Y. Optimisation of warp and weft tension on STB machines. Mag.dissertation. Tashkent, 2007.
66 Uzakov U.T., Rasulov H.Y., Rakhimkhodjaev S.S. Tensioning of weft yarns on projector machines.
67 . Rakhimkhodjaev S.S. et al. "Device for tensioning weft yarns on the weaving machine" Copyright Certificate No. 1687665 B.I., No. 40, 1991.
68 Rakhimkhodjaev S.S., Rasulov H.Y. Analytical studies of utochina tensioning on searchlight machines.
69 Rasulov H.Y., Rakhimkhodzhaev S.S. Studies of the tension of the weft on the shockless weft brake.
71 Rakhmonov O. T. Optimisation of parameters of structure and production of cotton-silk fabrics. Mag. dissertation. Tashkent, 2008.
72 .Lee E. A study of warp thread tension per fabric formation cycle. Mag.dissertation. Tashkent, 2003.
73 . Rasulov H.Y., Rakhimkhodjaev S.S., Kadyrova D.N. Tempering and tensioning of warp yarns on STB machines. Problems of textile. № 2, 2008.
74 Rakhimkhodzhaev S.S., Kadyrova D.N., Rasulov H.Y. Influence of medium parameters on movements of fabric down in elastic system of machine dressing. Problems of textile. № 2, 2014.
75 Rakhimkhodjaev S.S., Kadyrova D.N. Investigations of braking and feeding of weft on weaving machine. Problems of textile. № 4, 2002.
76 Muradova D.R., Umarova S.R., Rakhimkhodjaev S. S. Analytical investigations of movements of fabric edge on weaving machines. Scientific and practical conference of TITLP. Tashkent, 2016, pp. 101-104.
77 .Muradova D.R., Sobirova G.N., Rakhimkhodjaev S. C. Influence of the method of weft yarn laying on starting stripes in fabric. TITLP Scientific and Practical Conference. Tashkent, 2016, pp. 99-101.
78 . Muradova D.R., Rasulov X.Y., Rakhimkhodjaev S. S. Technological research of weaving machine JAT 810. Scientific and practical conference of TITLP. Tashkent, 2016, p. 73
79 Muradova D.R. Rakhimkhodzhaev S.S. Research of the movements of the fabric edge on weaving machines. Collection of Masters. TITLP. Tashkent, 2016.
80 . Rakhimkhodjaev S.S., Kadyrova D.N. New methods of measuring parameters of

weaving process. Problems of textile. № 3, 2002.
bb.Rakhimkhodjaev S.S. Improvement of warp tension regulation on shuttleless weaving machines during production of silk fabrics. Dissertation for the degree of Candidate of Technical Sciences. IvTI, Ivanovo, 1984.
72. Kadyrova D.N. Research and stabilisation of warp tension on needleless weaving machines. Dissertation, Tashkent, 2001.
73. Rakhimkhodjaev S.S. et al. Theory of tissue formation. Tashkent, 2007.
79 Sevostyanov A.G. Methods and means of research of mechanic-technological processes of textile industry.-M.: Legkaya Industriya, 1980. -392 c.
80 Ornatskaya V.A. et al. Designing and modernisation of weaving machines. M., L.I., 1986.
81 Rakhimkhodjaev S.S. et al. "Device for regulation of warp tension on a weaving machine" Copyright certificate No. 5014094. B.I.,No.40, 1997.
82 Kadyrova D.N. Investigation and stabilisation of warp tension on needleless weaving machines. Dissertation of Candidate of Technical Sciences, Tashkent, 2001.
83 Boimuratov B.H. Improvement of warp tempering and tensioning process on the weaving machine. Dissertation of Candidate of Technical Sciences, Tashkent. TITLP, 1998.
84 Martynova A.A. et al. Structure and design of fabrics. M., RIO MGTOA., 1999.
85 . Bogza A.D. et al. Investigation of reliability of weft laying process on STB machines, M. L.I., 1978.
86 . Rakhimkhodjaev S.S., Kadyrova D.N. Rheological properties of yarns in the elastic system of machine threading. Problems of textile. № 4, 2013.
87 Getsonok B.I. Statistical control of weaving process. M., 1983.
88 Bykadorov R.V. Regulation of fabric quality on weaving machines. M., L.I., 1984.

Printed by Books on Demand GmbH, Norderstedt / Germany